First Guide

to

MAMMALS

of North America

Peter Alden

Illustrated by
Richard P. Grossenheider

based on
*A Field Guide to the Mammals
of America North of Mexico*
by William H. Burt and
Richard P. Grossenheider

HOUGHTON
MIFFLIN
COMPANY

Library of Congress Cataloging-in-Publication Data

Alden, Peter.
Peterson first guide to mammals of North America.

Cover title: Peterson first guides. Mammals.
"Based on a field guide to the mammals of America north of Mexico, by William H. Burt and Richard P. Grossenheider."
Includes index.
1. Mammals—United States—Identification. 2. Mammals—Canada—Identification. I. Grossenheider, Richard Philip. II. Burt, William Henry, 1903–   . Field guide to the mammals. III. Title. IV. Title: First guide to mammals of North America. V. Title: Peterson first guides. Mammals.
QL715.A43   1987   599.0973   86-27821
ISBN 0-395-42767-3

Printed in Italy

NIL 10   9   8   7   6   5   4   3   2   1

Editor's Note

In 1934, my *Field Guide to the Birds* first saw the light of day. This book was designed so that live birds could be readily identified at a distance, by their patterns, shapes, and field marks, without resorting to the technical points specialists use to name species in the hand or in the specimen tray. The book introduced the "Peterson System," as it is now called, a visual system based on patternistic drawings with arrows to pinpoint the key field marks. The system is now used throughout the Peterson Field Guide Series, which has grown to over thirty volumes on a wide range of subjects, from ferns to fishes, rocks to stars, animal tracks to edible plants.

Even though Peterson Field Guides are intended for the novice as well as the expert, there are still many beginners who would like something simpler to start with—a smaller guide that would give them confidence. It is for this audience—those who perhaps recognize a crow or robin, buttercup or daisy, but little else—that the Peterson First Guides have been created. They offer a selection of the animals and plants you are most likely to see during your first forays afield. By narrowing the choices—and using the Peterson System—they make identification much easier. First Guides make it easy to get started in the field, and easy to graduate to the full-fledged Peterson Field Guides.

Roger Tory Peterson

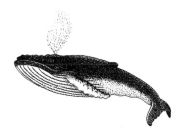

Introducing the Mammals

North America is rich with many different mammals living in many different habitats—forests, farms, prairies, deserts, swamps, mountains, oceans, and shores.

Many mammals are secretive and active only at night, so finding these species will take a great deal of patience, skill, and luck. But other mammals are active by day and are not at all shy. Some of the larger ones can best be seen in national parks and wildlife refuges. Many others can be seen—if you're alert and careful—in areas close to home in every state and province.

Mammal tracks are often easier to find than the animals themselves. Look for tracks in mud, sand, and snow and match them with the ones shown in this book.

Mammals are different from birds, reptiles, amphibians, and fish in a few important ways. Most mammals are covered with hair or fur, and almost all mammals have hair somewhere on their bodies. All mammals (with the exception of 3 very unusual species) are born live rather than hatched from eggs. All are warm-blooded, which means their body temperatures stay pretty much the same, regardless of the temperatures around them. And all mammals drink milk from their mothers' mammary glands when young.

Mammals are divided into smaller groups, depending on characteristics they have in common. The largest groups are called *orders.* The animals within each order are divided into smaller groups called *families.* For example, all mammals with special teeth for cutting meat belong to the Carnivore order. And the carnivores are divided into the Dog, Cat, Bear, Weasel, and Raccoon families.

The important characteristics of each order of mammals are described on the next few pages. Characteristics of some families are given throughout the book.

Note: Throughout the book, length is measured from the tip of the nose to the tip of the tail. For hoofed mammals (pp. 100–113), height is given instead, and is measured to the top of the shoulder.

MARSUPIALS

Marsupials have been on Earth longer than almost any other order. Their pea-sized young are born blind and naked and complete their development in a fur-lined pouch called a marsupium on the belly of their mother. While most marsupials are found in Australia and nearby islands, there are dozens of species in the Americas. One opossum species lives in N. America.

INSECTIVORES

These small, energetic mammals have long, pointed snouts and tiny, beadlike eyes. Unlike mice, most insectivores (such as shrews and moles) have short dense fur covering their ears and have 5 clawed toes on each foot. Using their numerous sharp teeth, insectivores prey on insects, earthworms, spiders, fish, frogs, and carrion (dead animals).

BATS

These are the only mammals that have wings and truly fly. (Flying squirrels and tropical mammals such as flying lemurs just glide.) A bat's wing is a tough membrane of skin that covers extra-long arm, hand, and finger bones. The membrane stretches from the forelimb, down the side of the body, to the leg. All North American bats use a form of "radar" called *echolocation* to catch their insect prey and to avoid objects. They give off high-pitched squeaks that strike objects and bounce back to their ears as echos. By listening to these echos, the bats can "read" their surroundings. Some bats roost in vast colonies in caves; others roost alone in trees or in attics, always hanging upside down. Some bats migrate south for the winter; others hibernate.

Bats are a major natural weapon against flying insect pests. Bat houses can be provided to encourage roosting in your backyard.

CARNIVORES pp. 22–51

A pet dog or cat can introduce you to the basic features of carnivores. Carnivores have long canine teeth, sharp "cheek" teeth for slicing meat, 5 toes on the front feet, and 4 or 5 on the hind feet. While some eat nothing but animals they have killed themselves, others eat a good amount of plant food as well as meat. North America is home to both the world's smallest and largest carnivores: the Least Weasel, at $\frac{1}{10}$ of a pound, and the Alaskan Brown Bear, which can reach over 1500 pounds.

SEALS pp. 52–57

Unlike whales and the other marine mammals on pp. 112–126, seals are closely related to carnivores. These ocean-dwelling mammals feed mostly on fish and other small aquatic animals. Their torpedo-like bodies feature front and hind flippers, and no tail is visible. Their nostrils close when submerged. Fine fur, which appears dark when wet but becomes paler when dry, covers most species of seals. Walruses have only a few sparse hairs. One seal—the Weddell—can dive to over 2000 feet and stay down for over 40 minutes. Seals and walruses "haul out" onto rocks, beaches, and ice floes only to rest, mate, and raise young.

RODENTS

pp. 58–91

These gnawing mammals outnumber those of all other orders combined, in both numbers of species and individuals. Rodents are active mostly at night and have bulbous eyes on the sides of the head. This helps them detect danger from nearly all directions at once. They are very active, and their high birth rate usually makes up for the great number of rodents that are caught by many predators. Most rodents have 4 toes on the forefeet (insectivores have 5) and 5 on the hind feet. They have 2 incisor teeth on top (rabbits have 4) and 2 below. Rodents lack canine teeth; they have a gap between the incisors and the molars.

RABBITS

pp. 92–99

Rabbits look a lot like rodents, but they have 4 upper incisor teeth rather than 2. There are 2 families in North America: the hares and rabbits in one and the pikas in another.

HOOFED MAMMALS

pp. 100–113

These heavy, plant-eating mammals have 2 toes on each foot. We have 4 families of hoofed mammals: the deer, the bovids (cattle), the pronghorns, and the peccaries. The first 3 families are *ruminants*. Their stomachs harbor bacteria and protozoa that aid in the digestion of plant fiber—a large part of these mammals' diets.

Harvest Mouse
and nest

EDENTATES pp. 112–113

These mammals live only in the Americas. This group includes the anteaters, sloths, and armadillos. One armadillo species lives in North America, while all other endentates live in Latin America. Anteaters have no teeth, but the others have many peglike teeth.

SIRENIANS pp. 112–113

These large, nearly hairless, cylinder-shaped mammals live in warm rivers and coastal waters of the tropics and subtropics. Their forelegs are in the form of flippers, their hind legs are absent, and their tail is broad and flat. Like whales, sirenians never leave the water. But unlike whales, they eat nothing but plants. The Manatee is the only North American sirenian.

WHALES pp. 114–126

Although fishlike in form, whales (including dolphins and porpoises) are true mammals. All whales have a pair of front flippers and a flat tail. They breathe through 1 or 2 nostrils on the top of the head, which are called blowholes. The cloud of vapor created as they exhale is called a blow, or spout, and is useful in spotting and identifying different species of whales. Most whales are fast swimmers and deep divers, reaching up to 23 mph and remaining submerged up to 2 hours.

Within these groups (orders), the sequence of species described and the names used are essentially the same as in *A Field Guide to the Mammals of America North of Mexico.* Although we have included all the more common and interesting North American mammals in this First Guide, these species are just a sampling. You will soon be ready for the full treatment found in that Field Guide, by William H. Burt and Richard P. Grossenheider.

MARSUPIALS: OPOSSUMS

OPOSSUM   To 40 in. long
                                          (including tail)*

This housecat-sized, nocturnal mammal is
whitish gray, with a *white face* and
pointed nose. Some Opossums (particularly
in the South) are blackish. All have thin,
round *black ears.* An Opossum can hang
from branches with its long, *rat-like tail,*
which is *pink* with a *black base.* The hind
feet have grasping "thumbs," which help
the Opossum grip branches and other
objects. Found on farms, in forests, and by
streams, the Opossum feeds on fruits, nuts,
bird eggs, insects, and carrion (dead animals). When frightened, an Opossum may
go into a state of shock and become so stiff
it appears to be dead. This behavior often
protects it from predators that attack only
live animals. Opossums are slowly expand-
ing their range from the Southeast and are
now also found throughout New England
and the Great Lakes region and along the
Pacific Coast. The dozen tiny young—each
the size of a lima bean—climb the mother's
long fur to her pouch immediately after
birth and stay there for another 2 months.

* length includes tail throughout book

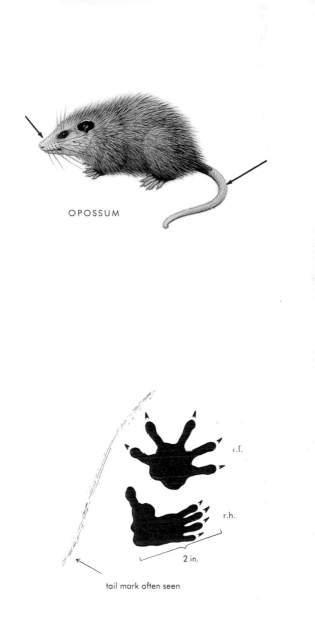

OPOSSUM

r.f.

r.h.

2 in.

tail mark often seen

INSECTIVORES: SHREWS

Shrew Family

Shrews look like moles but are slimmer and have visible eyes. They feed under leaves on the ground, rarely burrowing into the ground. Because they are extremely active day and night, they must eat more than their own weight in food daily. There are 30 species in North America.

LEAST SHREW   To 3¼ in. long
Tiny and *cinnamon-colored,* with a very *short* (¾-in.) *tail.* Found throughout south-eastern U.S., from Florida north to the Great Lakes and west to Texas and the Great Plains, chiefly in meadows and marshes.

MASKED SHREW   To 4½ in. long
Grayish brown, with a *long, bicolored tail* (brown above, *buff* below). Often the most common shrew in moist forests and brush in the northern U.S. and southern Canada.

ARCTIC SHREW   To 4⅔ in. long
Lives in bogs and among the leaf litter of northern forests. Has a *dark brown back* (nearly black in winter), *rusty sides,* and a *whitish belly.*

SHORTTAIL SHREW   To 5⅕ in. long
Large and *lead-colored.* It has *no visible ears* and a *short tail.* Attacks worms, snails, and even mice, killing them with its poisonous saliva. Found in a wide range of habitats east of the Great Plains.

NORTHERN WATER SHREW   To 6½ in. long
Large and *blackish gray,* with contrasting *silver underparts.* Can dive deeply and even run on the water after its prey. Found in the forests of New England, the Appalachians, and the Rockies north into Canada.

LEAST SHREW   MASKED SHREW

ARCTIC SHREW

SHORTTAIL SHREW

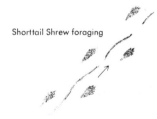

Shorttail Shrew foraging

NORTHERN WATER SHREW

INSECTIVORES: MOLES

Mole Family

Moles are stouter than shrews, with a *flexible, naked snout.* Their eyes are weak, but their senses of smell and touch are sharp. Moles have spade-like feet with *soles* that *turn outward.*

Low ridges meandering across lawns are a sign that moles have been moving just below the surface of the ground. They also "dive" deep into soft soils with a sort of breaststroke, sometimes digging at a speed of a foot per minute. They eat huge quantities of earthworms, grubs, slugs, insects, and occasional mice and tubers. The 7 North American species are absent from the Rockies and the Great Basin, where soils are too hard.

STARNOSE MOLE   To 8½ in. long
Dark brown to black, with a *star-shaped nose* of 22 tentacles. The hairy tail narrows near the body. Often seen above ground or swimming. It ranges from the Carolinas to Labrador and west to Minnesota.

HAIRYTAIL MOLE   To 7 in. long
A *slate-colored* mole with a *short, hairy tail.* Lives throughout the Appalachians, New England, and southern Canada.

EASTERN MOLE   To 8 in. long
The largest mole in the East and Great Plains. It has a *pink, naked, short tail.* The body varies from *slate-colored* in the North to *brown and golden* in the South and West.

TOWNSEND MOLE   To 9 in. long
Found only in the humid belt of northern California and the Pacific Northwest in moist meadows, gardens, and coniferous forests. As in the Eastern Mole, the front feet are *broader than long.* The tail is *slightly hairy.*

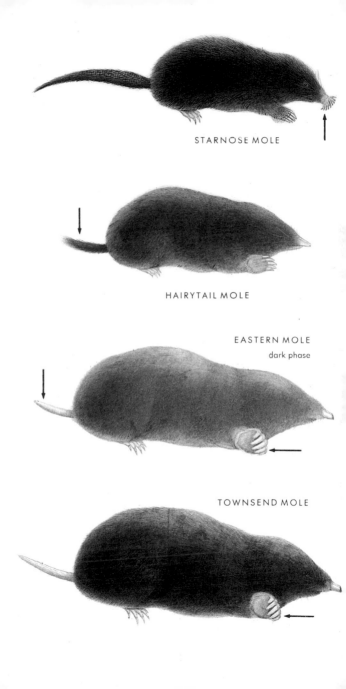

STARNOSE MOLE

HAIRYTAIL MOLE

EASTERN MOLE
dark phase

TOWNSEND MOLE

BATS

LEAFNOSE BAT   To 3½ in. long

A grayish bat with long ears. Like most members of its family, it has a *leaflike flap* of thick skin projecting upward from its nose. It roosts in caves and old mine tunnels in Arizona, southern California, and southern Nevada. Unlike most bats, it can hover and also swoop down and snatch large insects from the ground.

LONG-EARED MYOTIS   To 5½ in. long

Lives only in western coniferous forests, often near buildings. Has *large black ears* and long, glossy, *pale brown fur.*

LITTLE BROWN MYOTIS   To 5¼ in. long

Abundant in large colonies in all states and provinces, but not found along Gulf Coast. This bat is small and has medium-sized ears. The black hair has long, *glossy tips.* You may see it before sundown and recognize it by its erratic flight. Individuals that breed in the North migrate south for the winter.

BIG BROWN BAT   To 7 in. long

This common and widespread bat is large and has *all-black wings.* A fast flier (up to 40 mph), it feeds mainly on beetles. It is *dark brown* in moist areas and *paler brown* in deserts. This bat often winters in buildings and is usually solitary.

LEAFNOSE BAT

LONG-EARED MYOTIS

LITTLE BROWN MYOTIS

BIG BROWN BAT

BATS

SILVER-HAIRED BAT　　　　To 6 in. long
This bat is nearly black, but *silver-tipped
hairs* on the neck give it a *frosted* look. Flies
high and straight among trees in the middle and northern U.S. and Canada.
Migrates south for the winter.

WESTERN PIPISTREL　　　　To 5 in. long
Our smallest bat. It flies slowly (no more
than 6 mph) and erratically in arid areas of
the West and Southwest, often during the
day. It is *pale yellowish* or ashy gray.

EASTERN PIPISTREL　　　　To 5¼ in. long
One of the smallest eastern bats. It is *yellowish brown* to drab brown. It has a *slow,
erratic flight* and is often on the wing before
sundown.

RED BAT　　　　To 7¼ in. long
The sexes of this bat are of a different color:
males are *bright orange-red*, while females
are *dull red*. Both have a *frosted look*, with
hairs tipped with white. They roost alone on
tree branches and feed in pairs while flying
steadily and rapidly on a regular 100-yard
course. They are widespread except in
deserts and the Rockies, and they migrate
south for the winter.

SEMINOLE BAT　　　　To 6¼ in. long
This *mahogany brown* bat is found in the
southeastern U.S. from Pennsylvania to
Texas. It roosts low in trees and has habits
similar to those of the Red Bat.

HOARY BAT　　　　To 8½ in. long
This rarely seen bat lives throughout most
of the U.S. and Canada and is the largest
bat in the East. *White-tipped hairs* cover its
yellowish brown to *mahogany brown* body,
while the *throat is buffy*. Leaves late from
its daytime roost in conifers and flies high.
Migrates south for the winter.

SILVER-HAIRED BAT

WESTERN PIPISTREL

EASTERN PIPISTREL

female

male

RED BAT

SEMINOLE BAT

HOARY BAT

BATS

EASTERN YELLOW BAT   To 6½ in. long
A *pale yellowish brown* bat with long, silky
fur. It can be found in wooded areas of the
southeastern coastal states from Virginia to
Texas. Feeds mainly at medium altitudes
and often roosts in clumps of Spanish
moss.

EASTERN BIG-EARED BAT   To 6¼ in. long
This *pale brown* bat has *huge ears* joined
in the middle and *2 prominent lumps on
the nose.* It occurs throughout the Southeast and often hovers to pluck insects from
plants.

PALLID BAT   To 7 in. long
This bat lives in the western deserts. It is
the palest of those bats with *large ears.*
When disturbed, it gives off a skunk-like
odor. It feeds low to the ground, often pick-
ing up beetles, scorpions, and grasshoppers.

MEXICAN FREETAIL BAT   To 6¼ in. long
This bat and the next species are members
of a bat family with a *tail extending well
beyond* the edge of the *tail membrane.*
This *chocolate brown* bat is our smallest
"freetail" and is one of the commonest
mammals in the southern U.S., with a population of more than 100 million. During
the day it roosts in buildings and caves. In
Texas and at Carlsbad, New Mexico, vast
clouds of these bats exit from caves at dusk.
They return at dawn, after feeding up to
150 miles away on moths and other insects.

WESTERN MASTIFF BAT   To 10¾ in. long
The *largest* bat in North America. It is *chocolate brown* with a *long, free tail* and *enormous ears.* It is found in small colonies in
canyons and on cliffs along the Mexican
border and in southern California. Its loud
voice can be heard by humans.

EASTERN
YELLOW BAT

EASTERN
BIG-EARED BAT

PALLID BAT

MEXICAN
FREETAIL BAT

WESTERN
MASTIFF BAT

CARNIVORES: BEARS

Bear Family

Bears are the world's largest land-dwelling carnivores. They walk on the *entire foot*, rather than on the toes as cats and dogs do. They have short tails (usually not visible), small ears, and small eyes. Their eyesight is poor, but they have an excellent sense of smell. Most bears spend all winter in a den, where the female gives birth to her tiny cubs. Bears do not hibernate, but rather fall into a deep sleep from which they can awaken quickly.

BLACK BEAR   To 6 ft. long

Our *smallest* bear. In the East it is *nearly black* and in the West it is *black to cinnamon* or even *yellowish*. Its face is *roundish in profile* and always brown, and it usually has a white patch on its chest. Widespread in many forests, swamps, and mountain areas and is increasing in some parts of the East close to cities. Chiefly nocturnal, but it does forage in daytime, particularly in national parks. Eats berries, nuts, tubers, insects, small mammals, bird eggs, honey, carrion, and garbage. It can run up to 30 mph, and it weighs up to 475 pounds.

GRIZZLY BEAR   To 7 ft. long

Color varies from *pale yellowish* to *dark brown*. *White tips on its hairs* give it a grizzled appearance. Its face is *dish shaped* in profile, and it has a *distinct hump* at the shoulder, which is lacking in the Black Bear. Its front claws, which are longer than a Black Bear's, are useful in digging up rodents and excavating dens. Eats meat, fruit, grasses, fish, and other foods. Makes its own trails, which it uses over and over. At one time it was widespread in the Great Plains and the entire West. Now it survives mainly in national parks such as Yellowstone, Glacier, Banff, Jasper, and Denali. It can weigh up to 850 pounds.

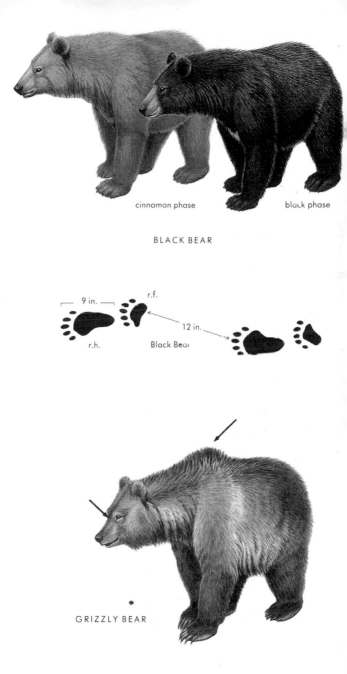

cinnamon phase

black phase

BLACK BEAR

9 in.

r.f.

12 in.

r.h.

Black Bear

GRIZZLY BEAR

CARNIVORES: BEARS

POLAR BEAR   To 8¾ ft. long

A *very large, white* bear, often with a yellowish tinge. Its *nose is black* and the eyes reflect a silvery blue. It lives on ice floes, barren shores, and tundra around Hudson's Bay, throughout the Canadian Arctic, and along the coast of Alaska south to Nome. A strong swimmer, it often is seen in open water. In late summer and autumn it is common around the town of Churchill, Manitoba, where it has become a tourist attraction. This bear hunts by sneaking up on seals resting on ice or by waiting for them at their breathing holes in the ice. It also feeds on dead whales, birds, bird eggs, and some vegetation, such as mushrooms, grasses, berries, and seaweed. The polar bear dens for winter in steep snowbanks and emerges in late March. Males are quite a bit larger than females and can weigh up to 1100 pounds.

ALASKAN BROWN BEAR   To 9 ft. long

This bear—the world's largest—is closely related to the Grizzly. It shares the Grizzly's *dish-faced profile* and the noticeable hump above the shoulder, but its *front claws are shorter,* and it is usually *darker brown,* with a yellowish tinge. The Alaskan Brown Bear ranges along the mountains, coasts, and islands of southern Alaska. Those on Kodiak Island are particularly large. This bear emerges from its den in spring to feed on seaweed and carrion, foraging both day and night. As summer comes, it grazes on grasses and sedges and then congregates on rivers during salmon runs. It eats many rodents and stranded whales. Unprovoked attacks on humans are rare.

POLAR BEAR

ALASKAN BROWN BEAR

CARNIVORES: RACCOONS

Raccoon Family

Raccoons are dog-sized mammals with long-ish tails that feature rings or bands. They eat a wide variety of animal and vegetable foods and often live in small family groups.

RACCOON To 40 in. long

Has a *black "bandit" mask* that shows up well on its whitish face. The body is brownish gray and the *tail has rings* of black and yellowish white. Mainly nocturnal, but will forage in daylight. Eats fruit, nuts, grain, insects, bird eggs, and many aquatic animals such as frogs, salamanders, crayfish, and fish. Found mainly in forests along streams and is increasing rapidly in urban areas, where it raids garbage cans. Does not hibernate, but will den up in very cold weather. Found throughout the U.S. and southern Canada, except for some places in the Rockies and deserts.

COATI To 50 in. long

A mostly tropical mammal, found in the U.S. only in southeastern Arizona, southwestern New Mexico, and along the Rio Grande in Texas—mostly in forested hill country. Has a *long snout* and a *long, weakly banded tail* that is often *carried high.* Active by day in bands of up to a dozen. May be seen on the ground or in trees. It roots grubs and tubers from the ground with the aid of its tough nose pad. Also eats fruit, nuts, lizards, scorpions, and tarantulas. Also called coatimundi.

RINGTAIL To 31 in. long

A nocturnal denizen of woods, brush, and rocky areas—particularly along streams—from Oregon to Texas and into the Southwest. It is pale yellowish gray, with *short legs* and a *long, bushy tail* that has many *black-and-white rings.* It eats mice, birds, insects, lizards, and some fruit.

RACCOON

4¼ in.

l.f. l.h.

Raccoon

COATI

3 in.

r.h. Coati r.f.

RINGTAIL

2⅝ in.

r.f. r.h.

Ringtail

CARNIVORES: WEASELS

Weasel Family

This large family includes weasels, badgers, skunks, otters, and minks. All are small to medium-sized carnivores with low bodies; longish tails; short legs; short, rounded ears; and thick, silky fur. Many have strong scent glands used for defense, protecting food caches, and social signaling. Males often are much larger than females.

WOLVERINE To 41 in. long

The largest member of the weasel family, found in the forests and tundra of northern and western Canada, Alaska, and in a few places in the Rockies and the Sierra Nevada. It weighs up to 60 pounds and looks like a small bear with a *heavy, bushy tail*. It has dark brown fur except for the yellowish temples and *broad yellow stripes* running from each shoulder to the base of the tail. This powerful and ferocious hunter preys on beavers, deer, porcupines, birds, and squirrels. It can kill prey as large as a moose or elk when the prey is bogged down in deep snow. A wolverine can drive bears and cougars from their kills and has a reputation for robbing traps and food caches of trappers. It marks any surplus supply of food with a smelly scent called musk, which repels other animals.

MINK To 26 in. long

Found throughout most of Canada and the U.S., except for dry areas of the Southwest. Prowls the shores of ponds, lakes, and streams, hunting small mammals, birds, fish, crayfish, and frogs. The mink is an *excellent swimmer* and will pursue prey in water. It is a *rich dark brown* with a *white chin patch* and has a slightly bushy tail. Since it is mostly active at night, it is not often seen. Its eyes glow yellowish green when light is shined on them. Males can weigh up to 3 pounds; females are smaller.

WOLVERINE

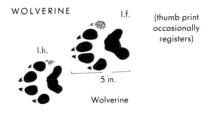

l.f.

l.h.

(thumb print occasionally registers)

5 in.

Wolverine

MINK

Mink

l.f.

r.f.

l.h.

1–2 ft.

r.h.

l.f.

r.f.

CARNIVORES: WEASELS

MARTEN To 26 in. long
About the size of a Mink, but with *yellow-ish brown* fur and a bushy tail. Its legs and tail are darker brown, and it has a *pale, buffy patch* on its throat and breast. Found in the northern forests of Canada, Alaska, the Rockies, the Pacific Northwest, the upper Great Lakes, the Adirondacks, and northern New England. The Marten is a quick and agile tree-climber, and it specializes in capturing squirrels—particularly Red Squirrels. It also hunts rabbits, mice, and birds and feeds on bird eggs, berries, seeds, and honey when it finds them. It is chiefly nocturnal, but can also be seen early and late in the day. Fur trapping and lumbering have wiped out the Marten in many areas, but it is now making a comeback.

FISHER To 40 in. long
The Fisher, weighing up to 12 pounds, is much larger than a Mink or Marten, and it lacks their pale chin and breast patches. It is *dark brownish black* overall, but white-tipped hairs give it a *frosted* appearance. The Fisher also is smaller and more slender than the Wolverine, with *no yellow stripes*. It lives in northern forests, sharing the range of the Marten, but it is absent from Alaska and the southern Rockies. It is active day and night both in trees and on the ground, but it is not as good a climber as the Marten. It preys on porcupines by flipping them over and attacking their undersides, which are unprotected by quills. It also feeds on snowshoe hares, squirrels, chipmunks, mice, fruit, and fern tips. Despite its name, the Fisher does not catch fish.

MARTEN

2 in. l.h.

Marten

l.f.

FISHER

3 in.

l.f.

Fisher

CARNIVORES: WEASELS

BADGER   To 28 in. long

The Badger has a distinctive *black-and-white face* and a *white stripe* from its nose to its shoulders. The wide, flat body is yellowish gray, becoming more *yellowish* on the tail and belly. It has short black legs with extremely *long front claws,* which it uses for digging rodents from the ground. It feeds on ground squirrels, gophers, rats, mice, birds, and even rattlesnakes. Its long hair protects it from snakebites, unless a snake strikes it directly on the nose. The Badger forages day and night but is more nocturnal where it is threatened by humans. It can weigh up to 25 pounds and defends itself well when cornered.

BLACK-FOOTED FERRET   To 24 in. long

At one time widespread in plains and high desert from Texas and Arizona north to Alberta, this ferret is now the most endangered mammal in North America. For millions of years it lived among the West's vast colonies of prairie dogs, feeding on the rodents as its main prey. But ranchers have poisoned almost every dogtown and thereby have driven the ferrets to near-extinction. At this writing a few pairs survive in captivity, and very small populations may still exist in Wyoming and elsewhere. This ferret has *yellowish buffy brown fur,* set off by *black feet* and a *black tail tip.* The face is white with a *black "bandit" mask.* In addition to feeding on prairie dogs, this ferret preys on ground squirrels, gophers, mice, birds, and small reptiles.

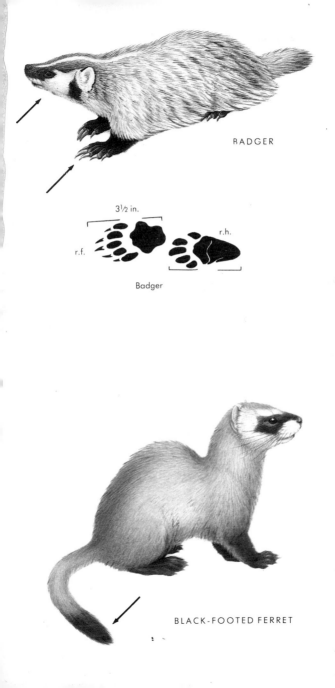

BADGER

3½ in.

r.f.   r.h.

Badger

BLACK-FOOTED FERRET

CARNIVORES: WEASELS

SHORTTAIL WEASEL (ERMINE)
To 13 in. long

In summer this weasel is dark brown above, with contrasting *white feet* and underparts and a white line down the inside of each hind leg. The Shorttail has a *black tip* on its tail even in winter, when the rest of its body becomes *pure white.* This winter color phase is called an Ermine. An expert mouser, the Shorttail also captures other small mammals and birds. It lives from Alaska and the Canadian Arctic south to Virginia, the Upper Midwest, the Rockies, and the Pacific Northwest, where it is light brown in winter rather than white. This weasel prefers bushy and wooded areas.

LEAST WEASEL
To 8 in. long

This is the *smallest carnivore* in the world. It is all brown above and *white below, including its feet.* Its *short tail* is brown. In the North it is *all white in winter,* It ranges from Alaska and Canada south to the northern Great Plains, Great Lakes, and the Appalachians, but is not found in the West, the South, or New England. It prefers meadows and fields but also is found in open woods. It feeds at night, almost entirely on mice.

LONGTAIL WEASEL
To 16½ in. long

The Longtail occurs in 48 states and southernmost Canada. It is the only weasel found in the Sun Belt. The Longtail superficially resembles the Shorttail in summer, but note the *longer, black-tipped tail;* the *yellowish white underparts;* and the *dark brown feet* (not white). Unlike the Shorttail, this weasel has no white line down the inside of each hind leg. In the North the Longtail becomes *white in winter* except for its black tail tip. In the Southwest a form with white patches on the face occurs. The Longtail feeds on mice, chipmunks, rats, and shrews, but can also catch prey larger than itself, such as rabbits.

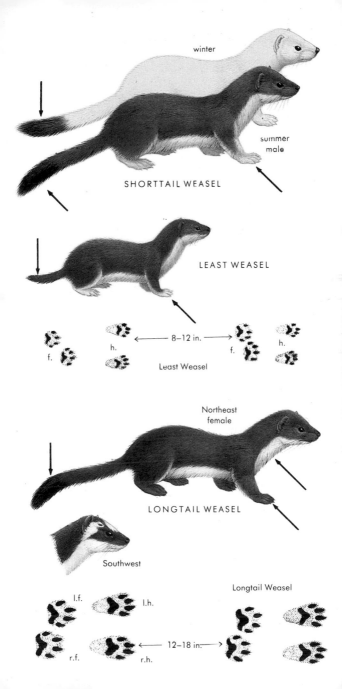

winter

summer male

SHORTTAIL WEASEL

LEAST WEASEL

←— 8–12 in. —→

f.   h.   f.   h.

Least Weasel

Northeast female

LONGTAIL WEASEL

Southwest

Longtail Weasel

l.f.   l.h.

r.f.   r.h.   ←— 12–18 in. —→

CARNIVORES: SKUNKS

Skunks are black with various white patterns and have a long, fluffy tail, which they often hold high over the back. When threatened, skunks spray a stinging, acrid, yellowish liquid called musk from a gland at the base of the tail. A skunk can spray accurately at distances up to 15 feet. Skunks are omnivorous—they eat snakes, lizards, insects, frogs, fish, bird eggs, small mammals, fruit, corn, and carrion.

STRIPED SKUNK   To 28 in. long
Our best-known skunk is found in 48 states and southern Canada. Note the white nape, the thin white stripe on its forehead, and the *2 broad white stripes* extending to its tail. Chiefly nocturnal, this skunk occurs in woods, prairies, suburbs, and even cities. It is active all year, but may den up in coldest weather.

SPOTTED SKUNK   To 22 in. long
Our *smallest skunk*, with an irregular pattern of *white spots and stripes* all over its black body. It is sometimes found in forested areas, where it may climb trees. But it prefers open, brushy areas (including deserts) west of the Mississippi and south of the Ohio River.

HOODED SKUNK   To 31 in. long
Found only along streams and on rocky ledges with thick brush, from southern Arizona and New Mexico to southwestern Texas. It has *2 thin white side stripes*, a shaggy *ruff on the neck*, and a *very long black tail*.

HOGNOSE SKUNK   To 31 in. long
This skunk has a *long, pig-like snout*. The entire *back and tail are pure white* and the underparts are solid black. The Hognose Skunk is found from the Gulf Coast of Texas west to the Colorado River. It has *oversized front claws*, which it uses for digging up insects, rodents, mollusks, snakes, and tubers.

STRIPED SKUNK

SPOTTED SKUNK

HOODED SKUNK

HOGNOSE SKUNK

CARNIVORES: OTTERS

RIVER OTTER   To 47 in. long
A large, weasel-like aquatic mammal. *Rich brown above,* with a *silvery sheen below.* Has webbed feet and a long tail with a *thick base.* Swims rapidly under water or on the surface, stopping now and then to look around by raising its head. Feeds on fish, frogs, crayfish, and muskrats. The River Otter is a playful, sociable animal whose presence is often noted by slides worn into streambanks and snowbanks. It often runs several miles over land to visit distant lakes and streams. Originally found over most of the continent, the River Otter was eliminated from many areas by overtrapping and loss of habitat. It is now protected and is making a slow comeback in many areas. Weighs up to 25 pounds.

SEA OTTER   To 49 in. long
Found along the Pacific Coast from central California to the Aleutians of Alaska. It is best seen from Point Lobos (south of Carmel) in California and on Alaskan cruises. It feeds, sleeps, mates, and gives birth at sea, coming ashore only during severe storms. The Sea Otter swims on its back and often uses a rock as a tool to break sea urchins and abalones on its chest. Note its long whiskers; *huge webbed, flipper-like feet;* and *yellowish gray face and neck.* It is otherwise dark brown, with white-tipped hairs that give it a frosted look. The Sea Otter nearly became extinct by the early 1900s due to uncontrolled killing for its thick, valuable fur. With protection it is now making a comeback in some areas. This huge otter can weigh up to 85 pounds.

l.h. 2¾ in.

River Otter

l.f.

River Otter
slide on snow
or mud bank—
about 8 in. wide

RIVER
OTTER

SEA OTTER

CARNIVORES: DOGS

Dog Family

The wolves, foxes, and coyotes have long,
narrow muzzles; erect, triangular ears;
long, slender legs; and bushy tails. Their
sense of smell is excellent, and they have
keen sight and hearing. Unlike cats, they
are unable to retract their claws.

GRAY WOLF · To 67 in. long

Our largest and most powerful dog, with
males weighing up to 120 pounds. This wolf
usually is a *grizzled gray* with a black-
tipped tail. Many individuals are nearly
white (in the Arctic) or *nearly black. Car-
ries its tail high* when running. Compared
to the Coyote, a wolf has *short, rounded
ears* and a broad face. It hunts both night
and day in family groups (packs) of up to a
dozen. A dominant male leads the pack, and
the rest of the group help care for the
young. Gray Wolves feed mainly on old,
weak, and diseased deer, caribou, and
moose, particularly those slowed by deep,
crusty snow. They also feed on rodents,
birds, fish, and berries. This wolf communi-
cates by a wide variety of howls and barks.
It once ranged throughout most of the continent, but now—due to relentless killing by
humans—survives in good numbers only in
Canada, Alaska, and Minnesota.

RED WOLF · To 57 in. long

A southern wolf with *reddish tawny legs,
muzzle, and ears.* Weighing up to 70
pounds, it is smaller and less powerful than
the Gray Wolf. The Red Wolf may be mis-
taken for the Coyote, but unlike a Coyote,
the wolf runs with its *tail held high.* It once
ranged from Pennsylvania to Texas, feeding
chiefly on rodents, birds, and crabs. Shooting, trapping, habitat loss, and inbreeding
with Coyotes have made it extinct in the
wild. Captive-bred Red Wolves are now
being reintroduced to a National Wildlife
Refuge in North Carolina.

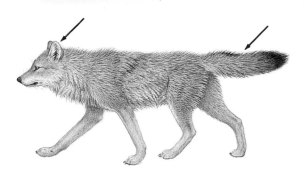

GRAY WOLF

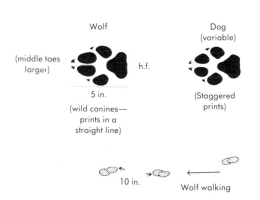

RED WOLF

Wolf

(middle toes larger)

h.f.

5 in.

(wild canines—
prints in a
straight line)

Dog
(variable)

(Staggered
prints)

10 in.

Wolf walking

CARNIVORES: DOGS

COYOTE   To 53 in. long
Larger than a fox but smaller than a wolf. It is gray or reddish gray, with *rusty legs, feet, and ears, and white underparts.* Unlike wolves, the Coyote runs with its *tail held down between its legs.* It also has *larger ears* than a wolf. The Coyote is active day and night and can be seen in most western wildlife refuges and parks. It usually hunts alone, for rabbits, hares, mice, ground squirrels, birds, frogs, and snakes. It can run up to 40 mph to catch faster prey. Occasionally, several Coyotes may down a larger animal such as a deer. Coyotes also feed on carrion (dead animals). Populations in the West are increasing, despite trapping, shooting, and poisoning by ranchers. Coyotes have adapted to suburbia in the West and have spread east to Massachusetts. A Coyote may weigh up to 50 pounds.

SWIFT FOX   To 32 in. long
A rare fox of the Great Plains, from Alberta to Texas. *Buffy yellow,* with a conspicuous *black tip on its tail.* It is smaller than a Coyote, and it lacks the white-tipped tail and black lower legs of the Red Fox. It feeds on rodents and insects, but unfortunately has been eliminated from many areas by poisoning and habitat loss. The similar **Kit Fox** (not shown) lives in deserts west of the Rockies, from Oregon to Mexico.

ARCTIC FOX   To 31 in. long
Found only beyond treeline in northern Canada and western Alaska. Like many other arctic mammals, it has *short, rounded ears* and heavily furred feet. Occurs in two color phases—the blue phase and white phase. In summer both types are brownish slate with no white tail tip. In winter some become *all white,* others *slate blue.* The Arctic Fox follows Polar Bears for scraps; scavenges on carcasses of marine mammals; captures lemmings, hares, and birds; and forages for berries.

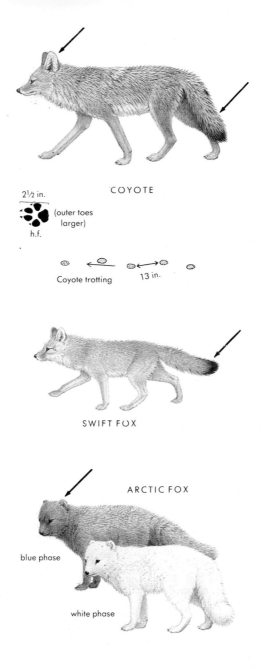

COYOTE

2½ in.
(outer toes larger)
h.f.

Coyote trotting ← → 13 in.

SWIFT FOX

ARCTIC FOX

blue phase

white phase

CARNIVORES: DOGS

RED FOX To 41 in. long
The classic quarry of the foxhunter, the wily Red Fox is *reddish yellow* with *black legs.* Its bushy tail is a mixture of black and reddish hairs; the *tail tip is white.* A rare color variation is the black (or silver) phase, which retains the white tail tip. The cross phase resembles a Gray Fox but has a dark cross over the shoulders and down the middle of the back. The Red Fox is active in open country and forests almost everywhere, but is absent from much of the Pacific Coast, southwestern deserts, and the Rockies. It feeds on insects, birds, rodents, rabbits, berries, and fruit—usually at night. Weighs up to 15 pounds.

GRAY FOX To 45 in. long
Note the *pepper-and-salt gray coat,* with *rusty patches* on the neck, belly, legs, and tail. The muzzle is blackish, contrasting with the white cheeks and throat. The tail does *not* have a white tip; a black stripe down its entire length ends in a *black tip.* The Gray Fox inhabits the eastern, southern, and southwestern U.S., barely reaching Canada. It hunts in chaparral, open forests, and rocky areas. It specializes in capturing rodents, plus some insects, fruit, acorns, birds, and eggs. This fox regularly forages *in trees,* and will climb trees to escape enemies. The **Insular Gray Fox** (not shown) inhabits 6 of the Channel Islands off the coast of California.

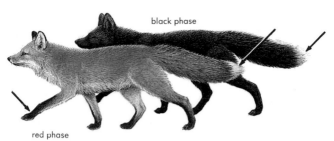

black phase

red phase

RED FOX

1¾ in.

h.f.

Red Fox

GRAY FOX

11 in.

Gray Fox trotting

CARNIVORES: CATS

Cat Family

Cats have shorter faces than most dogs do, with smaller ears. Unlike dogs, they have *retractile claws*. Their tails, whether short or long, are relatively thin, not bushy. Their vision is excellent. Most hunt at night.

MOUNTAIN LION   To 90 in. long

Many prefer to call this the Cougar or Puma. It is large and long, with a yellowish tawny or grayish coat. Its *long tail* is *tipped with dark brown*, as are the backs of the ears and sides of the face. While the young are spotted all over, the adults are unspotted. The Cougar survives mostly in forests, semi-arid areas, and mountains of the West. The only Cougars known east of the Mississippi live in southern Florida—a few individuals, known as the endangered "Florida Panther." The Cougar feeds mainly on deer, but it also hunts hares, rodents, Coyotes, Raccoons, and occasionally domestic animals. Chiefly nocturnal, it hunts by stalking its prey on the ground and by ambushing from trees. Males can weigh up to 200 pounds.

JAGUAR   To 84 in. long

This powerful cat ranged throughout much of the southern parts of California, Arizona, New Mexico, and Texas until the early 1900s. Its back and sides are tawny and evenly covered with *black spots* in the form of *rosettes* (circles of spots with a spot in the center). The belly is white with single black spots. The Jaguar is chiefly nocturnal and is rarely seen, even in Latin America. It hunts peccaries, rodents, and birds, and may kill cattle occasionally. Weighs up to 250 pounds.

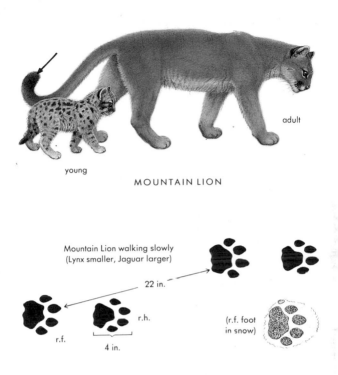

young

adult

MOUNTAIN LION

Mountain Lion walking slowly
(Lynx smaller, Jaguar larger)

22 in.

r.h.

(r.f. foot
in snow)

r.f.

4 in.

JAGUAR

CARNIVORES: TROPICAL CATS

OCELOT   To 50 in. long
The Ocelot has *rows* of blackish spots with
dark centers, making it appear *striped*. It is
much more heavily spotted than a Bobcat
and has a much longer tail. The Ocelot is
found in southern Texas and southeastern
Arizona. It inhabits thick thornscrub, rocky
areas, and woods. It hunts chiefly at night,
and preys on rabbits, birds, fish, frogs, and
rodents. It is adept at hunting in trees, in
water, and on the ground. This cat suffered
a huge decline in numbers due to the fur
trade, pet trade, and loss of habitat. The
sale of kittens and importation of pelts have
been banned in the U.S. Weighs up to 40
pounds.

JAGUARUNDI   To 54 in. long
Another tropical cat, found primarily in
Latin America but may be seen on rare
occasions in southeastern Arizona and in
southernmost Texas along the Rio Grande.
Note its extremely *long, thin tail* and very
short legs, similar to those of a weasel. This
wild cat is twice as big as a house cat. It
occurs in 2 color phases—all gray, or reddish. The Jaguarundi lives in bushy areas,
thorn thickets, and mesquite. It is active in
twilight and after dark. Feeds on rats, mice,
rabbits, and birds. This cat swims well and
also catches fish.

OCELOT

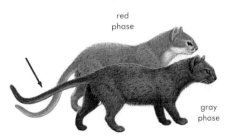

red
phase

gray
phase

JAGUARUNDI CAT

CARNIVORES: SHORT-TAILED CATS

LYNX   To 40 in. long

A *bob-tailed* cat of northern forests. Note the *long tufts* on the ears, the conspicuous facial ruff, and the *black tail tip.* The feet are *huge* and act as snowshoes, enabling this cat to walk over deep snow as it stalks and chases Snowshoe Hares, rodents, and birds. When populations of Snowshoe Hares crash—as they do naturally every 10 years or so—the Lynx also suffers a decline. It is found throughout much of Canada and Alaska and is also called the Canadian Lynx. The Lynx can also be seen in the upper Great Lakes region, Maine, the Rockies, and the Cascades. Trappers eagerly seek its luxurious, long, soft, grayish buff fur, which is mottled with brown. The Lynx is usually nocturnal, but it is forced to hunt by daylight in the far North during the long days of summer. Weighs up to 30 pounds.

BOBCAT   To 35 in. long

A southern cousin of the Lynx, the Bobcat is often called the Bay Lynx or "wildcat." In some areas its range overlaps that of the Lynx. The Bobcat can be recognized by its *dark spots* (particularly on the legs), its smaller ears, and the tip of its tail, which is black *only on top.* Its color varies from a warm tawny brown in summer to grayish in winter. This cat hunts birds and mammals, such as rabbits, hares, mice, squirrels, and porcupines, at night. It lives in forests, scrub, swamps, rocky country, and some farm lands, from the Canadian border southward to Mexico. The Bobcat has been heavily trapped for its fur, and is absent now in most of the Midwest. It is vocal at times, and can make a rich variety of coughs, yowls, and screams.

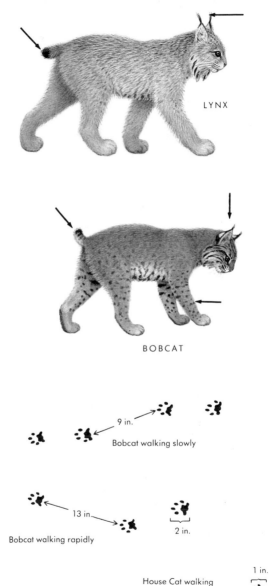

LYNX

BOBCAT

9 in.

Bobcat walking slowly

13 in.

Bobcat walking rapidly

2 in.

1 in.

House Cat walking
8 in. to next track

SEALS: EARED SEALS

Eared Seal Family

These seals have small *external ears*. Their hind flippers can be *turned forward*, which enables these seals to get about on land. Males are up to 4½ times larger than females.

NORTHERN SEA LION   To 10½ ft. long

An enormous sea lion—males weigh up to 2000 pounds, females up to 600 pounds. This sea lion has *yellowish brown fur* and a *low forehead.* Usually quiet when not molested, but it can give a lion-like roar. This seal lives on the *Pacific* Coast, from the Alaskan islands and bays south to California. It is also called Steller's Sea Lion.

CALIFORNIA SEA LION   To 8 ft. long

Smaller and darker than the Northern Sea Lion—males weigh up to 600 pounds, females up to 200 pounds. This sea lion has a *high forehead.* It is extremely noisy, barking constantly. This is the "seal" most often seen performing in aquarium shows and circus acts. Females and young remain along the California coast near their rookeries (breeding grounds), while males migrate northward, to British Columbia. This sea lion is a fast swimmer, capable of reaching 25 mph.

ALASKA FUR SEAL   To 6 ft. long

Males (which weigh up to 600 pounds) are *blackish* above, *reddish* on the belly, and have a *brownish face.* Females (which weigh up to 135 pounds) are gray above and reddish below. This fur seal breeds chiefly on the Pribilof Islands of Alaska and migrates as far south as California during the winter. Note the relatively *short face.* The **Guadalupe Fur Seal** (not shown), which has a *longer snout,* can be seen at San Nicolas Island, California.

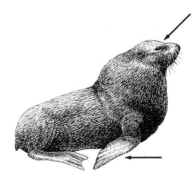

NORTHERN SEA LION
8–10½ ft.
(2.4–3.2 m)

**CALIFORNIA
SEA LION**
5½–8 ft.
(1.7–2.4 m)

**ALASKA
FUR SEAL**
4–6 ft.
(1.2–1.8 m)

WALRUS AND EARLESS SEALS

WALRUS   To 12 ft. long
Like the eared seals, the walrus has hind
feet (flippers) that can be turned forward for
walking. However, like the earless seals, the
Walrus has no external ears. Both males
and females have *2 large white tusks* projecting downward from the upper jaw. The
Walrus's thick, *hairless* skin appears black
when wet, pinkish when dry. The Walrus
dives down to the sea bottom to feed on
mollusks and crustaceans, which it detects
with its facial bristles. It sometimes catches
small seals and will feed on dead whales.
Found along coasts in the far North—from
the Alaska peninsula to eastern Canada.
Males weigh up to 2700 pounds.

Earless Seal Family
These seals have only an opening in the
head for ears—*no visible ears.* Their hind
flippers cannot be turned forward, so on
land they are limited to wriggling. In all species except the Elephant Seal both sexes are
roughly the *same size.*

ELEPHANT SEAL   To 20 ft. long
The world's *largest* seal—males (bulls)
weigh up to 8000 pounds (4 tons). Females
weigh up to 2000 pounds (only 1 ton).
Found along the Pacific Coast from California to the Gulf of Alaska. Most easily seen at
Año Nuevo Point near Santa Cruz, California. This seal is essentially hairless, like a
Walrus. Its color ranges from brown to grayish. Older males have *large, overhanging
snouts,* which can be inflated to amplify
their loud bellowing. The Elephant Seal
sleeps on beaches by day and feeds on
sharks, squid, and rays at night.

BEARDED SEAL   To 10 ft. long
A large (up to 875 pounds), dark grayish to
yellowish seal, with *long bristles* on each
side of the muzzle. Found in arctic waters
only as far south as western Alaska, Hudson's Bay, and Labrador. The **Gray Seal**
(not shown), a large, grayish or black seal
with a long face, replaces the Bearded Seal
from Labrador south to Nantucket.

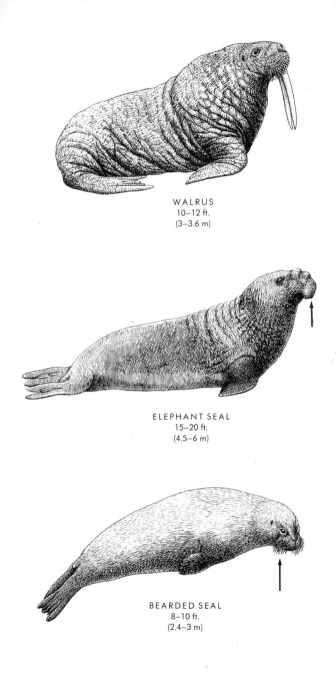

WALRUS
10–12 ft.
(3–3.6 m)

ELEPHANT SEAL
15–20 ft.
(4.5–6 m)

BEARDED SEAL
8–10 ft.
(2.4–3 m)

SEALS: EARLESS SEALS

HARBOR SEAL To 5 ft. long
The common small seal of harbors, river mouths, and rocky coasts from the Carolinas and California north into the Arctic and Hudson's Bay. Varies in color from brown to gray, often with many *small spots*. This seal hauls out on beaches and rocky shores at low tide and feeds on fish, squid, and octupus on the incoming tide. The **Ringed Seal** (not shown) is similar but has many *pale circles (rings)* on its dark back. It is found on arctic coasts from Alaska to Labrador.

RIBBON SEAL To 5½ ft. long
Inhabits ice floes and open waters of the Bering Sea off western Alaska north of the Aleutians. The males are brownish, with a beautiful pattern of *wide, creamy or buff rings* around the front flippers, neck, and rump. Females are grayer, with indistinct rings.

HARP SEAL To 6 ft. long
The male is a striking *yellowish white,* with a *dark brown head* and variable *dark patterns* on the back, with a few random spots elsewhere. The female is less distinctly marked or has no dark markings. The young are yellowish white all over, with large, round black eyes. Harp Seals are found from the Maritimes north to the eastern Canadian Arctic. They migrate up to 6000 miles a year to Baffin Island and the Grand Banks of Newfoundland. The young are born on the edge of the ice pack in the Gulf of St. Lawrence. Each year, a limited number of young are killed for their pelts.

HOODED SEAL To 11 ft. long
A large gray seal with numerous *large black blotches.* The male has an inflatable "bag" on top of his head which can be blown up to make him look more formidable. Found only in the northwest Atlantic, from the Gulf of St. Lawrence to Baffin Island.

HARBOR SEAL
5 ft.
(1.5 m)

RIBBON SEAL
5 ft.
(1.5 m)

HARP SEAL
6 ft.
(1.8 m)

HOODED SEAL
7–11 ft.
(2.1–3.3 m)

LARGE RODENTS

BEAVER
To 40 in. long

Our largest rodent (weighs up to 60 pounds). It has rich brown fur, huge gnawing teeth, webbed hind feet, and a naked, scaly, *paddle-shaped tail*. It fells trees by gnawing and uses branches to build stick and mud dams. The dams create protective ponds, in which it builds a domed lodge, or house. The Beaver may live along rivers, where it burrows into banks and does not build dams. It feeds on bark and small twigs, stashing a supply of branches underwater for winter use. At one time found almost everywhere in North America, the Beaver has survived heavy trapping and is being reintroduced into many areas, including western rangelands.

PORCUPINE
To 31 in. long

Our only mammal with *long sharp quills*, which cover all but the belly. *Heavy-bodied* and *short-legged*, it is most often seen as a black ball resting high in a tree, where it eats bark, buds, and small twigs. Active chiefly at night. The Porcupine can cause damage to trees and wooden buildings and poles, especially in areas where its predators have been wiped out. It lives in forests of Alaska, Canada, the West, the upper Great Lakes, and the Northeast.

APLODONTIA
To 18 in. long

The world's most primitive rodent. Looks like a plump, *tailless* Muskrat with small eyes and ears. Found mostly in the moist forests of the Pacific Coast, from San Francisco Bay and the Sierra Nevada north through western Oregon, Washington, and southern British Columbia. Makes extensive runways and burrows beneath dense streamside vegetation. The Aplodontia is active mostly at night, when it emerges to feed on herbaceous plants and on shrubs. Also known as Mountain Beaver.

BEAVER

Beaver

3–6 in.
About 4 in.
between tracks.
Hind track
covers front.

Tree cut by Beaver

PORCUPINE

r.f.

r.h.

3 in.

Porcupine

APLODONTIA

AQUATIC RODENTS

MUSKRAT   To 25 in. long
The Muskrat has dense, rich brown fur and
a silver belly. It features a long, scaly, naked
tail that is flattened from side to side
(rather than top to bottom, as in the Beaver, which is much larger). Muskrats use
aquatic plants in marshes to build conspicuous *conical houses* that rise up to 3 ft.
above the waterline. They also live in lakes
and streams, where they dig burrows with
underwater entrances into banks. A Muskrat marks its territory by leaving musky
secretions on vegetation. It feeds on cat-
tails, sedges, rushes, water lilies, frogs,
clams, and fish (rarely). It is often seen and
is frequently trapped for its valuable fur
over most of the U.S. and Canada except for
the far North, Florida, and most of the
Southwest. Weighs up to 4 pounds.

FLORIDA WATER RAT   To 15 in. long
Also known as the Round-tailed Muskrat,
this water rat *lacks the flattened tail* of the
northern Muskrat. It replaces its larger rela-
tive in Florida and the Okeefenokee area of
Georgia. The Florida Water Rat builds *bulky
nests* out of tightly woven sedges at bases of
stumps and mangroves as well as in savan-
nas near streams. It feeds on water plants
and crayfish in bogs, marshes, lakes, and
everglades. Weighs only 12 ounces.

NUTRIA   To 42 in. long
This large aquatic rodent weighs up to 25
pounds. A native of Argentina and Brazil, it
escaped from Louisiana fur farms in the
1940s and has replaced Muskrats in much
of the South and Pacific Northwest. It is
grayer than the Muskrat, with a *longer,
round tail*. It nests among marsh plants or
digs a burrow in streambanks above the
waterline.

Muskrat houses in marsh

MUSKRAT

r.h.   r.f.

3 in.   Muskrat walking   Tail mark

FLORIDA WATER RAT

NUTRIA

RODENTS: MARMOTS

Marmots are *oversized squirrels,* with *short legs* and long digging claws on the front feet. All hibernate during winter.

WOODCHUCK (GROUNDHOG)   To 27 in. long
This marmot is found throughout most of Canada and the northeastern and midwestern states. It is a uniform *frosted brown,* with *dark feet* and some white around the nose. It digs a den up to 5 ft. deep and up to 30 ft. long. Woodchucks feed on plants such as grasses, clover, and alfalfa. In some areas they are a nuisance in gardens and corn fields. Active day or night, they often can be seen swimming, resting in trees, and feeding beside roadways. The Woodchuck differs from the Beaver and Muskrat by its medium-length bushy tail. Its call is a shrill whistle. Weighs up to 10 pounds.

YELLOWBELLY MARMOT   To 28 in. long
Replaces the Woodchuck in the interior western U.S., from New Mexico to British Columbia and from the Black Hills to California. It is a rich *yellowish brown,* with a yellow belly and bushy tail, which is reddish with a black tip. The *black head* has *white patches* in front of the eyes and a rusty patch below each ear. This marmot feeds on grasses and forbs in mountains and valleys, usually near rocky areas. It gives *high-pitched chirps* from a lookout boulder when alarmed.

HOARY MARMOT   To 31 in. long
A very large marmot (weighs up to 20 pounds). *Silvery gray,* with a *black-and-white* head and shoulders. It inhabits rockslides near meadows in mountains of Alaska and western Canada south to Idaho. Gives a *shrill whistle.* Males may be seen engaged in stand-up wrestling matches. Three very similar marmots (not shown) are found only on Vancouver Island, the Olympic Peninsula, and northern Alaska.

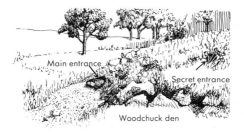

Main entrance

Secret entrance

Woodchuck den

1½ in.

l.h.   l.f.

WOODCHUCK

YELLOWBELLY
MARMOT

HOARY
MARMOT

RODENTS: PRAIRIE DOGS

The western plains at one time swarmed with billions of these chunky, broad-headed squirrels. Both species have very *short, hairy tails;* short legs; and strong claws. These social animals live in "towns" and post sentinels to warn the colony of incoming coyotes, badgers, ferrets, snakes, and birds of prey. They *bark* (like a dog) and bob up and down in excitement before retreating below. Their underground villages can be 16 ft. deep and can extend another 20 ft. on the level, with side chambers for storage and nesting and with escape tunnels. Prairie dogs feed on grasses, roots, and blossoms. In the past, ranchers have shot, poisoned, trapped, and gassed them to keep them from making new burrows, fearing that livestock would break a leg in the prairie dogs' entrance holes. Grain farmers of today have eliminated most of the rest. Those that survive are found in parks and other uncultivated areas.

BLACKTAIL PRAIRIE DOG   To 21 in. long
This prairie dog ranges across the *shortgrass* prairies of the western Great Plains from Montana to western Texas. It weighs up to 3 pounds and can be identified by its *black-tipped tail.* A large colony can be seen at Wichita Mountain National Wildlife Refuge, in Oklahoma.

WHITETAIL PRAIRIE DOG   To 14½ in. long
Found in upland meadows and brushy country with scattered junipers and pines, at altitudes of 5,000–12,000 ft. Ranges farther west than the Blacktail, into Wyoming, Utah, Arizona, western Colorado, and western New Mexico. Has a *shorter tail* than the Blacktail, with a *white tip.*

BLACKTAIL PRAIRIE DOG

WHITETAIL PRAIRIE DOG

Prairie dog town

RODENTS: GROUND SQUIRRELS

A variable group of small to medium-sized, ground-dwelling squirrels found in the western half of the continent and on the tundra. Many sit up like prairie dogs, but ground squirrels have longer faces or longer tails, or both. Eight of the 17 species are shown here. (Consult a *Field Guide to the Mammals* for the details on the rest.)

CALIFORNIA GROUND SQUIRREL To 20 in. long

Brown with *buffy bands* on rear half. Dark back contrasts with *pale gray neck* and shoulders. The white-fringed tail is much less bushy than that of the Western Gray Squirrel. This ground squirrel prefers pastures, grain fields, lightly treed slopes, and rocky ridges. It eats a wide variety of green vegetation, fruits, mushrooms, acorns, seeds, berries, bird eggs, and insects. Its burrows are up to *200 ft. long.* Found in more humid areas of California and western Oregon.

ROCK SQUIRREL To 21 in. long

The *largest* ground squirrel of the Southwest, ranging from Great Salt Lake and the Colorado Rockies south to the Mexican border of Arizona and western Texas. A slightly *mottled gray* squirrel with a somewhat *bushy tail.* Some individuals are darker on the head and back. Inhabits rocky canyons and slopes with boulders.

COLUMBIAN GROUND SQUIRREL To 17 in. long

Rather large, with a short, bushy tail, *mottled gray upperparts,* and *reddish legs* and *face.* Lives in meadows and at forest edges from the Canadian Rockies south to Idaho.

ARCTIC GROUND SQUIRREL To 21 in. long

A large squirrel of Alaska and northern Canada, west of Hudson's Bay. Found in tundra and brushy mountain meadows. The only ground squirrel in its range. *Dusky brown* with many *pale spots.* Very vocal—sounds include a *sik-sik* call.

CALIFORNIA
GROUND
SQUIRREL

ROCK
SQUIRREL

COLUMBIAN
GROUND
SQUIRREL

ARCTIC
GROUND
SQUIRREL

RODENTS: GROUND SQUIRRELS

THIRTEEN-LINED GROUND SQUIRREL
To 11½ in. long

Unique *dark brown stripes* above, with *buffy spots* within them. This ground squirrel originally lived in the shortgrass prairies from Alberta to Texas, but—with the clearing of forests—has spread eastward to Ohio and Michigan. It is often seen feeding on seeds and insects along roadsides and on lawns and golf courses.

SPOTTED GROUND SQUIRREL
To 9½ in. long

A pale brownish squirrel, with *light buffy spots* on its back. The tail is pencil-like, not bushy. A shy resident of sandy forests, brush, and grassy parks from Nevada to the Rio Grande and eastern Arizona. In southwestern Texas and southeastern New Mexico, look for the **Mexican Ground Squirrel** (not shown), which is similar but *darker brown*, with *whiter spots* in distinct *rows*.

FRANKLIN GROUND SQUIRREL
To 16 in. long

Found in the *Midwest* and northern Great Plains, from Missouri to Saskatchewan. The *largest* and *darkest* ground squirrel in its range. Dark gray with tawny overtones and a fairly long tail. Will climb trees, but usually seen on the ground. The **Uinta Ground Squirrel** (not shown) is similar (drab gray and brown), but has a *cinnamon-colored face.* It is seen near lodges at Yellowstone and Grand Teton.

GOLDEN-MANTLED SQUIRREL
To 12½ in. long

A ground squirrel that looks like a chipmunk. Familiar to campers and visitors in most western parks, as it becomes tame due to hand-outs. Note its *copper-red head and shoulders,* "chipmunk" stripes (only on the back), and relatively small tail. Found in evergreen forests, in chaparral, and above timberline.

THIRTEEN-LINED
GROUND
SQUIRREL

SPOTTED
GROUND
SQUIRREL

FRANKLIN
GROUND
SQUIRREL

GOLDEN MANTLED
SQUIRREL

RODENTS: CHIPMUNKS AND ANTELOPE SQUIRRELS

Chipmunks

Alert, small, ground-dwelling squirrels with a slightly bushy tail that is flicked often. Note their narrow, *erect ears* and *cheek pouches,* which are frequently stuffed with food. All 16 species feature 5 dark and 4 paler *stripes down the back* and more or less distinct *white stripes* above and below the eye. Chipmunks feed on seeds, nuts, fruit, insects, and bird eggs.

Chipmunks are territorial and often chase each other to defend their burrows or food caches. A burrow can be 30 ft. long with side chambers, escape hatches, and an entrance by a boulder or stump.

TOWNSEND CHIPMUNK   To 12½ in. long
Large and dark brown. Found in the Pacific Northwest and Sierra Nevada.

EASTERN CHIPMUNK   To 10 in. long
Reddish rump. Eastern U.S. and Canada.

MERRIAM CHIPMUNK   To 12 in. long
Stripes are dark brown (not black) and gray. Central and southern California.

LEAST CHIPMUNK   To 9 in. long
Smallest and most variable in color. Stripes extend to *base of tail.* Ontario to Yukon, and interior western U.S.

CLIFF CHIPMUNK   To 10 in. long
Grayish with *indistinct stripes.* Southwest.

COLORADO CHIPMUNK   To 9½ in. long
Large *white patch* behind each ear. Upper Colorado Basin and southern Rockies.

Antelope Squirrels

Four similar species live in the *deserts* of the Southwest and Great Basin. Each has a *white stripe* on each side of back. All run with the *tail arched over the back.* The **Yuma** of southern Arizona is replaced by the **Whitetail** (not shown) between the Rockies and Sierra Nevada.

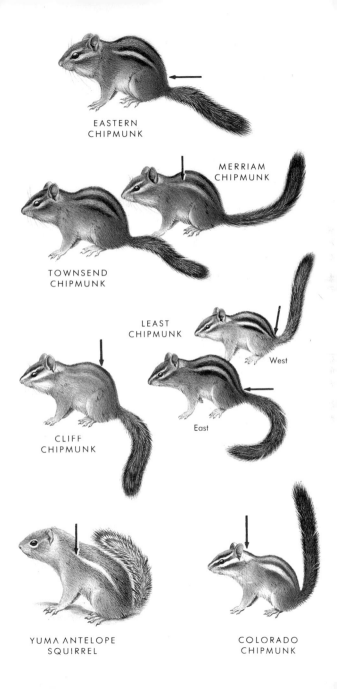

EASTERN
CHIPMUNK

MERRIAM
CHIPMUNK

TOWNSEND
CHIPMUNK

LEAST
CHIPMUNK

West

East

CLIFF
CHIPMUNK

YUMA ANTELOPE
SQUIRREL

COLORADO
CHIPMUNK

RODENTS: TREE SQUIRRELS (EASTERN)

EASTERN GRAY SQUIRREL   To 20 in. long
Perhaps the most familiar mammal of eastern North America. It is abundant in city parks, suburbs, and rural woodlands where there are plenty of nut trees. Its tail is very bushy and is bordered with white hairs. The body is *gray* with white underparts, and in summer it may appear tawny. Black and all-white individuals are not unusual. This squirrel feeds on a great variety of nuts, seeds, fungi, and fruits. It can destroy small trees by stripping the bark to reach sap. It stores nuts and acorns in the ground, many of which sprout. Nests in natural cavities and builds leafy nests in tree branches. As it dashes among tree branches, its tail helps it balance. It becomes a pest by stealing seeds people put out to feed birds, and by gnawing its way into buildings to spend the winter. Barks sharply or chatters when excited.

FOX SQUIRREL   To 29 in. long
The largest tree squirrel. Widespread in eastern U.S. and westward through the Great Plains, but absent northeast of Philadelphia. Normally it is *rusty* yellowish gray with a pale *yellowish orange belly*, but populations in the Chesapeake Bay region are often *pure gray* (no tawny) with white around the nose. In the Southeast it often has a *black head* with white on the nose and ears. Its habits, food, and calls are similar to those of the Eastern Gray Squirrel, but it can weigh twice as much (up to 3 pounds).

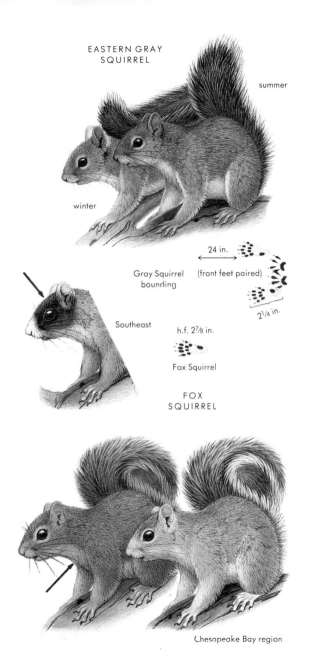

EASTERN GRAY
SQUIRREL

summer

winter

Gray Squirrel
bounding

24 in.

(front feet paired)

2¼ in.

Southeast

h.f. 2⅞ in.

Fox Squirrel

FOX
SQUIRREL

Chesapeake Bay region

RODENTS: TREE SQUIRRELS (WESTERN)

WESTERN GRAY SQUIRREL   To 24 in. long
A large, gray squirrel with a long, bushy tail and white belly. Lives in California, western Oregon, and Washington, in oak and pine-oak forests. Differs from the Eastern Gray (which has been introduced near Seattle and Vancouver) by its *dusky feet,* the reddish fur on the back of the ears, light banding on the underside of the tail, and lack of tawny overtones in summer. Feeds on acorns and evergreen seeds and thrives on walnut and almond plantations.

TASSEL-EARED SQUIRREL   To 21 in. long
America's most attractive and colorful tree squirrel. Gray, with a *chestnut back* and chestnut on the back of its *tasseled ears.* Note the contrast between *dark sides* and white underparts. The tail is usually gray above, *white below.* A tourist attraction in yellow pine forests of Arizona, New Mexico, Colorado, and southeastern Utah. Feeds on pine seeds, pinyon nuts, mistletoe, and inner tree bark. An isolated population lives on the north rim of the Grand Canyon. These individuals have an *all-white tail* and black (not white) underparts; they are known as the Kaibab Squirrel.

CHICKAREE   To 12 in. long
Also known as the Douglas Squirrel. This small tree squirrel replaces the Red Squirrel (p. 76) in the humid evergreen forests of the Pacific Northwest, from coastal British Columbia south to the Sierra Nevada. The *tail is blacker* than that of the Red Squirrel. *Dark gray* in summer; in winter it is paler gray, with *gray ear tufts.* Its habits are similar to those of the Red Squirrel.

WESTERN GRAY
SQUIRREL

south of
Grand Canyon

TASSEL-EARED
SQUIRREL

north rim of Grand Canyon
(Kaibab Squirrel)

CHICKAREE

summer

winter

RED SQUIRREL

summer

winter

Red Squirrel
h.f. 1¾ in.

RODENTS: SMALL TREE SQUIRRELS

RED SQUIRREL   To 14 in. long
Considerably smaller than the Eastern or
Western Gray Squirrel. This *rusty red*
squirrel has distinct summer and winter
coats, though it is always *white below*, with
a *red tail.* In summer it is a *darker* reddish
gray, with *black* side stripes. In winter it is
a *paler* rusty gray, with *rusty ear tufts* and
no side stripes. The Red Squirrel lives in
pine, spruce, and hardwood forests of
Alaska, Canada, the Rocky Mountains
(where it is called the Pine Squirrel), and
the northeastern U.S. south to the Smokies.
It feeds on seed, nuts, and fungi. Active all
year, and very vocal.

Flying Squirrels
Small, *nocturnal* squirrels with *large* black
eyes. *Gray-brown* to cinnamon above, white
below. These squirrels have broad folds of
skin that connect their front and back feet,
allowing them to glide down (not fly) from

Leaf nest of Tree Squirrel

Flying Squirrel gliding from den-tree hole

one tree to another. The flat, broad tail helps break their falls. Flying squirrels spend their daylight hours in natural and manmade cavities, emerging to feed on seeds, nuts, and insects. They utter faint, birdlike *chip* notes.

SOUTHERN FLYING SQUIRREL

To 10 in. long

Found in the eastern U.S. and southern Ontario. Olive-brown above and pure white below. The **Northern Flying Squirrel** (not shown) is slightly larger (up to 12 in. long). It is found in the western U.S. and much of Canada; its range overlaps that of the Southern Flying Squirrel in the Great Lakes and New England southward through the Appalachians.

SOUTHERN
FLYING
SQUIRREL

RODENTS: NATIVE MICE

WHITE-FOOTED MOUSE   To 8 in. long
Widespread in forests and brushy areas
from New England to Arizona and Montana.
Note its *large ears* and rich reddish brown
upperparts, contrasting with the pure
white feet and underparts. The tail is *usually no longer than the body.* This mouse
lives in abandoned bird nests, outbuildings,
and other animals' burrows. Active year
'round, it feeds on seeds and insects.

DEER MOUSE   To 9 in. long
Our most wide-ranging native mouse. Has a
bicolored tail and its body *color varies* from
grayish buff to deep reddish brown. Occurs
throughout Canada and U.S. except for
Alaska and the Southeast. The larger woodland form nests in hollow logs; the prairie
form digs small burrows. Feeds on seeds,
nuts, centipedes, and insects.

GOLDEN MOUSE   To 7½ in. long
A handsome, *golden-cinnamon* mouse with
white belly. Lives only in the *southeastern*
U.S., among *trees,* vines and brush. This
mouse feeds on insects and seeds of poison
ivy, sumac, greenbrier, and wild cherry.

JUMPING MICE   To 10 in. long
Four species of very *long-tailed* mice with
large hind feet and *small ears* edged with
white or buff. They are widespread except in
deserts and the Southeast. The **Woodland
Jumping Mouse** lives in the Northeast and
is active at night, while the **Meadow Jumping Mouse** (not shown) is often seen by day.

HARVEST MICE   To 6¼ in. long
Five species of small brown mice that
resemble the House Mouse. Active day and
night, they live in dense vegetation and
build large grass nests (see p. 8). These
mice feed on vegetation, seeds, and fruit.
They are found in the southeastern and
western U.S.

WHITE-FOOTED MOUSE

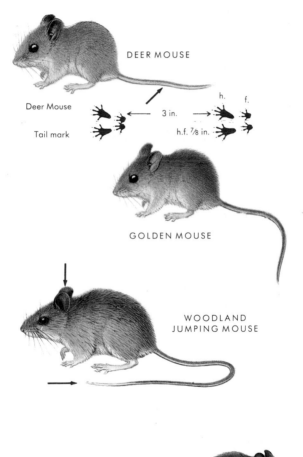

DEER MOUSE

Deer Mouse

Tail mark

h.   f.

← 3 in. →

h.f. ⅞ in.

GOLDEN MOUSE

WOODLAND
JUMPING MOUSE

EASTERN HARVEST MOUSE

RODENTS: NATIVE MICE

GRASSHOPPER MICE   To 7 in. long
Pale cinnamon or *pale gray* above, with
white underparts and a *short white tail*.
These stout, largely carnivorous mice prey
on other mice, grasshoppers, scorpions,
beetles, and lizards. Occasionally they eat
seeds. These mice live in burrows of other
animals and appear at night in open sandy
or gravelly sage and grasslands. The **Northern Grasshopper Mouse** occupies the Great
Plains, Great Basin, and western plateaus,
while the smaller **Southern Grasshopper
Mouse** (not shown) replaces it in southwestern deserts.

PINYON MOUSE   To 9 in. long
This mouse has even *larger ears* than the
Brush Mouse. The tail is bicolored and is
often slightly *shorter* than the body. The
Pinyon Mouse lives in rocky terrain with
scattered pinyon pines and juniper.

BRUSH MOUSE   To 8 in. long
A *large-eared*, gray-brown mouse, with
tawny sides and a *well-haired tail* that is
often slightly *longer* than the body. This
mouse lives in chaparral and brush of semi-
arid areas from Arkansas west to California.

HISPID COTTON RAT   To 14 in. long
Widespread in the South, from the Carolinas to the Rio Grande. Note its *long,
coarse fur*, heavily mixed with buff and
black hairs, and its *tail*, which is *shorter*
than the body. Lives in moist tall grass and
feeds on vegetation and bird eggs.

RICE RAT   To 12 in. long
Grayish brown above, with a pale gray or
buff belly. The *scaly tail* is *longer* than the
body and the feet are *whitish*. Inhabits mar-
shy areas of southeastern states. Active
chiefly at night, building or traveling on
runways through grass. This rat feeds on
seeds, vegetation, crabs, snails, and
insects. It swims well.

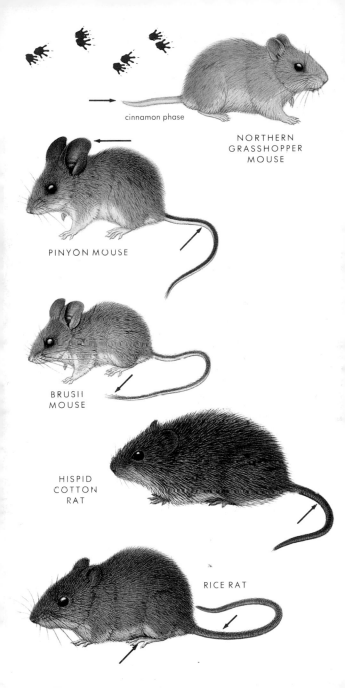

cinnamon phase

NORTHERN
GRASSHOPPER
MOUSE

PINYON MOUSE

BRUSII
MOUSE

HISPID
COTTON
RAT

RICE RAT

OPEN-COUNTRY RODENTS

Kangaroo Rats

Fourteen species of nocturnal rodents that live in arid areas of the western U.S. They have extremely *long, white hind feet,* and *powerful thighs* (crossed by a *white line*) that allow them to jump up to 6 ft. in one bound. The front feet are tiny by comparison. The dark tails with *white side stripes* and a *bushy tip* are much longer than the body. Size varies and body color varies from a pale sandy color to dark brown. Kangaroo rats feed on seeds and desert foliage.

BANNERTAIL   To 15 in. long
KANGAROO RAT
Boldly marked, with a *black-and-white tail.* Found in New Mexico, west Texas, and southern Arizona.

ORD KANGAROO RAT   To 10½ in. long
Great Plains and Great Basin from Canada to Mexico.

MERRIAM KANGAROO RAT   To 10 in. long
Our smallest species. California north to Nevada, east to Texas.

PLAINS POCKET GOPHER   To 13 in. long
One of a dozen species of pocket gophers living in the southern and western U.S. and Canadian prairie provinces. These rodents are named for their *fur-lined cheek pouches* (pockets). Note their large, yellowish, *exposed incisor teeth; heavily clawed front feet;* and a *short, naked, rat-like tail.* Pocket gophers make fan-shaped mounds when they dig their burrows. They are active day and night, but rarely leave their burrows. They feed on roots, tubers, and field crops.

CALIFORNIA POCKET MOUSE   To 9 in. long
One of the 20 species of small, nocturnal desert and plains mice with *cheek pouches* like those of a gopher and legs and feet similar to those of kangaroo rats. They are unpatterned gray or brownish above, and have long, thin tails.

82

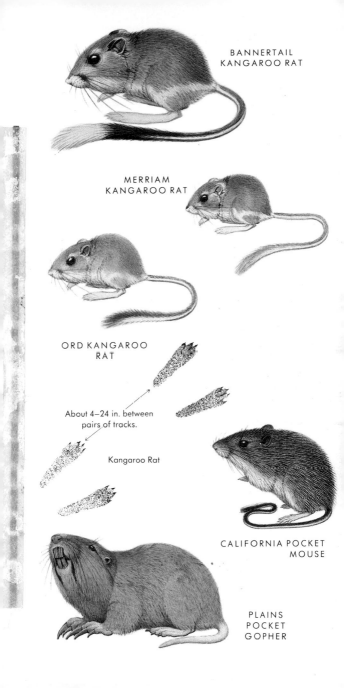

BANNERTAIL
KANGAROO RAT

MERRIAM
KANGAROO RAT

ORD KANGAROO
RAT

About 4–24 in. between
pairs of tracks.

Kangaroo Rat

CALIFORNIA POCKET
MOUSE

PLAINS
POCKET
GOPHER

RODENTS: VOLES

Redback Voles
BOREAL REDBACK VOLE To 6½ in. long
A small, relatively tame inhabitant of cool,
moist forests and bogs. Uses natural runways along logs, rocks, and trails. This
vole's *reddish back* contrasts with its gray
sides, although those in the Northeast are
all gray. Widespread in northeastern U.S.,
upper Midwest, Rockies, and Canada. The
California Redback Vole (not shown) of the
Pacific Northwest has dark olive-brown
sides and a dark chestnut back.

Other Voles
Active day and night, these rodents of
grassy areas dig underground burrows and
make runways up to 2 in. wide through
matted grasses. In winter, the round
entrances of their tunnels can be seen in
snow. Voles feed mainly on plants and
sometimes eat seeds or bark from tree roots
as well as green vegetation. The 17 species
are mainly brownish gray with long fur.
They have *small ears*, a tail that is *shorter*
than the body, and small, beady eyes.
MEADOW VOLE To 7½ in. long
The most widely distributed vole in North
America. Dark brown in the East to grayish
brown in the West. Found in *meadows* and
thick vegetation *near water*, from the
northern U.S. through Canada to Alaska.
YELLOWNOSE VOLE To 6½ in. long
Grayish brown, with a distinct *yellow
patch* behind the nose. Inhabits cool,
moist, rocky woodlands from eastern Canada south to the Smokies.
PRAIRIE VOLE To 6½ in. long
Widespread in prairies and dry areas from
the Ohio Valley to the Rockies. Common
along fencerows, railroad rights-of-way, and
old cemeteries.
PINE VOLE To 5 in. long
Widespread in eastern and southeastern
U.S., chiefly in broadleaf forests (despite the
name). This vole is a rich *auburn* color. It
has tiny ears, and a very *short tail*.

BOREAL
REDBACK
VOLE

MEADOW VOLE

YELLOWNOSE VOLE

PRAIRIE VOLE

PINE VOLE

RODENTS: PHENACOMYS AND LEMMINGS

TREE PHENACOMYS To 7 in. long
Bright reddish brown, with a *long, blackish, well-haired tail.* Lives high in spruce, hemlock, and fir trees, feeding on their needles. Builds huge twig nests up to 150 ft. above ground. Found only in northwestern California and western Oregon.

MOUNTAIN PHENACOMYS To 6 in. long
Found in grassy areas near mountain tops, rocky slopes, and coniferous forests in Canada and southwards in our western mountains. A *grayish brown* vole with white feet and a relatively short tail.

Lemmings

Five species of small, vole-like mammals, found chiefly in the tundra of the *far north.* Long, soft fur often conceals their tiny ears. Tail *very short.*

SOUTHERN BOG LEMMING To 6 in. long
A round, brown lemming that is *almost tailless.* It makes runways just below the ground in grassy meadows of the Midwest, northeastern U.S., and eastern Canada. Feeds on grasses and clover. The **Northern Bog Lemming** (not shown) is slightly larger. It lives in bogs, mountain meadows, and tundra from northern New England west to Alaska.

BROWN LEMMING To 6¾ in. long
Looks like an oversized vole with a grayish head and shoulders and a bright *reddish brown back and rump.* Does not turn white in winter. Ranges from Hudson's Bay west through Alaska and south to British Columbia, chiefly in tundra.

GREENLAND COLLARED To 6⅜ in. long
LEMMING
The distinctly *patterned* summer coat of this lemming includes a *black stripe* from head to tail on the back, a pale face, and a rusty collar. *All white in winter.* Restricted to *tundra* from Hudson's Bay west to Alaska Peninsula. The similar **Hudson's Bay Collared Lemming** (not shown) lives in northern Quebec and Labrador.

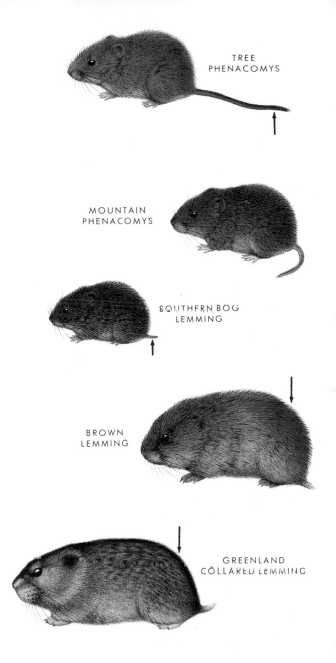

TREE
PHENACOMYS

MOUNTAIN
PHENACOMYS

SOUTHERN BOG
LEMMING

BROWN
LEMMING

GREENLAND
COLLARED LEMMING

RODENTS: WOODRATS

Also known as "pack rats" or "trade rats." These rats carry small objects—including pieces of rubbish, buttons, coins, and jack-knives—to their nests and store them there. Woodrats are about the same size as Old World (non-native) rats, but have a *hairy* (not scaly) *tail*, fine soft fur, larger ears, and *white* feet and underparts. The 8 species build a variety of huge *stick nests* in trees, cliffs, cacti, and brush. They are generally nocturnal and shy, but may be active and quite bold by day in caves and abandoned mine tunnels. Their food consists of insects, snails, buds, leaves, fruit, and fungi.

EASTERN WOODRAT   To 17 in. long
A large, *grayish brown* woodrat, darker than the others. Ranges through southeastern U.S. and southern plains, living in cliffs in the Appalachians and wooded swamps in the South.

WHITETHROAT WOODRAT   To 15½ in. long
Found in *deserts* of the southwestern U.S., chiefly in Arizona, New Mexico, and west Texas. Feeds on cactus, mesquite beans, and various seeds. Builds nests in extremely thorny cholla or prickly pear cactus.

DESERT WOODRAT   To 13½ in. long
Smaller and *paler* than the whitethroat. Found in deserts and rocky slopes of California, Nevada, Utah, and western Arizona. Builds nests on ground or along cliffs.

BUSHYTAIL WOODRAT   To 17 in. long
Grayish tawny to almost black above; easily identified by its long, *bushy, squirrel-like* tail. Widespread in pines and rocks of western mountains from Grand Canyon and Black Hills to Yukon.

EASTERN
WOODRAT

WHITETHROAT
WOODRAT

DESERT
WOODRAT

BUSHYTAIL
WOODRAT

OLD WORLD RODENTS

These invaders from Asia and Europe have adapted well to people's habits and habitats. They live in our buildings and ships, steal our stored food, and thrive on our garbage-rich society. Great nuisances and carriers of disease, these rodents can be told from our native mice and rats by their *long, hairless tails.*

HOUSE MOUSE   To 7 in. long

A small, *grayish brown* mouse. Lacks the contrasting white underparts of most native mice. Note its *scaly tail*, about as long as the body. This mouse lives in houses and other buildings throughout settled North America; it is occasionally found in cultivated fields.

NORWAY RAT   To 18 in. long

Also known as the Brown, Sewer, or House Rat. It originated in central Asia, spread across Europe between 1500 and 1700, and was brought to North America by ship around 1776. Grayish *brown* above, with gray (not white) underparts. Its long, scaly, naked tail is *shorter* than its body. Feeds on garbage, insects, stored grain, and will kill chickens. An excellent digger, making its own tunnels under buildings and dumps. Found throughout the lower 48 states and southern Canada.

BLACK RAT   To 18 in. long

Also known as the Ship or Roof Rat. Invaded Europe from southeast Asia centuries before the Norway Rat; brought to North America in the early 1600s. Displaced by the larger, more aggressive Norway Rat in many areas. Found chiefly in seaports, the southern U.S., and on the roofs of buildings. Two color phases—*brown* and *black*. The tail is *longer* than the body.

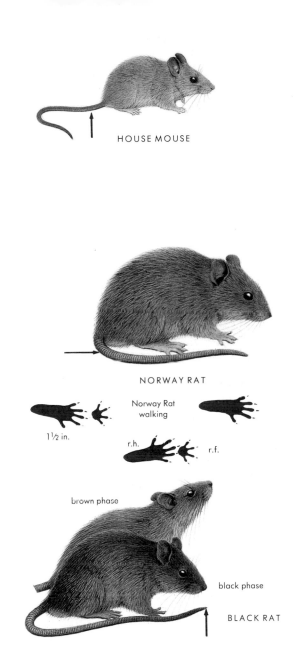

HOUSE MOUSE

NORWAY RAT

Norway Rat
walking

1 1/2 in.

r.h.   r.f.

brown phase

black phase

BLACK RAT

RABBITS: HARES

Hare and Rabbit Family

In comparison with the pikas (p. 98), hares and rabbits have long ears, long hind limbs, short "cottony" tails, and bulging eyes. In this group of mammals the females are larger than males. Although hares and rabbits are usually silent, they may squeal and often thump the ground with their hind legs to communicate danger.

Hares, including the misnamed "jackrabbits," are larger than rabbits and have longer ears. Their more powerful hind limbs enable them to leap distances of up to 20 ft. and run at speeds up to 35 mph. They make no nests, and their young can hop about within hours of birth.

WHITETAIL JACKRABBIT   To 26 in. long
Lives in open, grassy (or sagebrush) plains and barren fields of the northern U.S. and the Canadian prairies, from Wisconsin and Kansas west to the Cascades and the Sierra Nevada. *Brownish gray* in summer, with a *white tail.* In winter it is often *white* or pale gray, with black-tipped ears. Chiefly nocturnal, it feeds on grasses in summer and on buds, bark, and twigs in winter.

BLACKTAIL JACKRABBIT   To 25 in. long
Common in open prairies, fields, and deserts of southwestern U.S. north to Washington and South Dakota, including most of California and Texas. Grayish brown with large, black-tipped ears. The *top of the tail* and rump are *black.*

ANTELOPE JACKRABBIT   To 26 in. long
Dark brown back and rump *contrasts* with dazzling white sides. The enormous 8-inch ears lack black tips. Found in saguaro forests and cactus flats of southern Arizona. Feeds on cactus, coarse grasses, and seeds from late afternoon to midmorning.

WHITETAIL
JACKRABBIT

winter

summer

BLACKTAIL
JACKRABBIT

ANTELOPE
JACKRABBIT

Jackrabbit

l.h.

l.f.

←— 7–12 ft.—→

2¾ in. +

RABBITS: NORTHERN HARES

SNOWSHOE HARE   To 26 in. long
Dark brown in summer, including the tail,
becoming *white* with black-tipped ears in
winter. Also called the Varying Hare. In
spring and autumn it often appears
"patchy." It is found in northern forests,
swamps, and thickets of Alaska and Canada
southward to the Great Lakes, Rockies, and
Appalachians. Does not hibernate. Survives
the winter by feeding on buds, twigs, bark,
and frozen carrion. Its wide feet act as
"snowshoes." Known for its boom and bust
population cycles.

ARCTIC HARE   To 26 in. long
This hare lives on the arctic tundra of far
northern Canada, ranging south to Hudson's Bay and Newfoundland. *Grayish
brown* in summer, it turns *white* in winter
(except for its black-tipped ears). This hare
differs from the Snowshoe Hare by its *white
tail* in summer and its larger size. On Baffin
Island it remains white all year. Active year
'round and often found in large groups. It
often stands on its hind feet and hops with
its forefeet held against its chest. The **Tundra** or **Alaskan Hare** (not shown) is similar.
It lives in tundra and thickets of northern
and western Alaska.

EUROPEAN HARE   To 27 in. long
A non-native hare that now lives in the wild
in the northeastern U.S. and southern
Ontario. Larger than any other hare or rabbit in its range. Inhabits open fields in hilly
country. It escapes predators by making
wild, zigzagging dashes and doubling back
on its tracks. The body is brownish in summer and grayish in winter, and the *tail is
black above.*

r.h.
r.f.
Snowshoe Hare
6 in.
1–10 ft.

winter

summer

SNOWSHOE
HARE

winter

ARCTIC
HARE

summer

EUROPEAN
HARE

EASTERN RABBITS

Rabbits are smaller than hares, with shorter ears and hind legs. Although they are good runners, they usually try to hide in thickets to escape enemies. They make nests of their own fur, grasses, and leaves, where they raise their young. The young are born naked with their eyes closed and need weeks of care.

EASTERN COTTONTAIL To 19 in. long
Grayish brown mixed with black hairs above and a *rusty nape.* Top of feet *whitish* and the cottony tail is *white.* Widespread from the Atlantic Coast west to the Rocky Mountains and Arizona, in heavy brush, mixed woods and fields, swamp edges, and weed patches. It can damage gardens, shrubs, and small trees. The **New England Cottontail** (not shown) inhabits mountains in the eastern U.S. It usually *lacks the rusty nape,* has a *black patch* between the ears, and is *redder.*

MARSH RABBIT To 18 in. long
Restricted to southeastern bottom lands, swamps, and hummocks from coastal Virginia through Florida. It is *dark brown* with *small feet,* which are *reddish brown* above. Its tail is *small,* inconspicuous, and dingy white. Chiefly nocturnal, it searches for food such as bulbs, cane, and grasses in marshes. Escapes predators by leaping into water and swimming with only its eyes and nose above surface.

SWAMP RABBIT To 20 in. long
Our *largest* cottontail. Replaces the Marsh Rabbit from Georgia west to Texas and north to southern Illinois. Brownish gray, *mottled* with black above, it is larger than the Marsh Rabbit and has *paler rust* fur on the tops of its feet. The Eastern Cottontail has whiter hind feet and a more distinct rusty nape. The Swamp Rabbit is found in swamps, marshes, and bottom lands and is often seen swimming in water. It forages in water and ashore on cane and crops.

EASTERN
COTTONTAIL

l.f.   l.h.

1–7 ft.

Cottontail

MARSH RABBIT

SWAMP RABBIT

WESTERN RABBITS AND PIKA

DESERT COTTONTAIL   To 16 in. long
The common cottontail of low valleys, open
plains, and foothills from the Western Great
Plains through the arid Southwest to the
Pacific in California. The body is *pale gray
washed with yellow.* The ears are relatively
long. This cottontail can climb sloping
trees.

MOUNTAIN COTTONTAIL   To 25 in. long
Paler than the Eastern Cottontail, and *grayish* in color. Its *black-tipped ears* are
shorter than those of the Desert Cottontail.
The *only* cottontail in much of the mountains of the West from the Sierra Nevada
and New Mexico north to Alberta. Inhabits
thickets, sagebrush, rocky areas, and cliffs.

BRUSH RABBIT   To 14 in. long
A small, *dark brown* rabbit with relatively
small ears and *tail.* Lives in *chaparral* and
brush of California and western Oregon
only.

PYGMY RABBIT   To 11 in. long
The smallest North American rabbit. *Slate-
gray* with *pinkish tinge.* The *tail* is *nearly
hidden* and the *ears* are *small.* Digs its
own burrows among clumps of tall sagebrush in cooler deserts of the Great Basin.

PIKA   To 8½ in. long
Pikas are in a *separate family* from the rabbits and hares. They have *short, broad,
rounded ears* and *no visible tail.* They
occur in *colonies* on open rocky hillsides
high in our western mountains. Their call
is a *bleat* that is very hard to trace. Look for
their piles of *fresh hay* drying in the sun (in
winter, it will be stored under rocks). The
Collared Pika (not shown) of Alaska,
Yukon, and northern British Columbia has
a pale gray collar and a white belly.

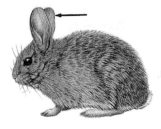

DESERT COTTONTAIL

MOUNTAIN COTTONTAIL

BRUSH RABBIT

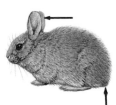

PYGMY RABBIT

PIKA

HOOFED MAMMALS: DEER
Deer Family

The members of this family have *2 toes* on each foot. The males have *antlers*, which begin to grow each spring as soft, tender bone covered with thin skin and fuzzy hair. Later in the year, after the antlers have hardened, the skin dries up and falls off. Antlers are used in courtship battles in the autumn, after which they drop off.

ELK (WAPITI)   To 5 ft. high (at shoulder)

Large deer with a *pale brown yellowish rump*, *short tail*, and *dark brown* legs, belly, neck, and head. Males have a *shaggy neck mane* and, in late summer and autumn, a set of large, spreading antlers. Usually seen in *groups* of 25 or more, with old males in separate groups in summer and both sexes together in winter. Feed in semi-open forests and mountain meadows in summer, descending to valleys in winter. Males bugle and battle for control of "harems" of females. The Elk has been killed off over much of its range in the East and the Great Plains and now survives chiefly in the Rockies and Northwest. Best seen in Grand Teton, Yellowstone, Olympic, Glacier, Rocky Mountain, Banff, and Jasper parks. Males weigh up to 1000 pounds.

WOODLAND CARIBOU   To 4 ft. high

Brown, shaggy fur and a *whitish neck and mane*. Males and most of the *females* grow antlers. The males' antlers are *large* and *partly flattened*. Both sexes have hooves that spread out in summer, making it easier to walk in bogs; in winter the toes close up, which gives this caribou a better grip on the ice. Although it has been killed off in much of its former range in the northern U.S., it survives in wilder Canadian forests such as Prince Albert Park, Saskatchewan. Males weigh up to 600 pounds.

BARREN GROUND CARIBOU   To 4 ft. high

Pale whitish, with a brownish wash on back. The *antlers are less flattened* than in the Woodland Caribou. Huge numbers migrate over the tundra of Alaska and the Canadian Arctic. Best seen at Denali Park in Alaska. Males weigh up to 400 pounds.

2–3 ft. to
next track.

Elk

4½ in.

ELK

Caribou

20–40 in. to next track.

4 in.

WOODLAND
CARIBOU

BARREN
GROUND
CARIBOU

HOOFED MAMMALS: DEER

WHITETAIL DEER   To 3½ ft. high (at shoulder)

Widespread in woodlands, swamps, and brush over most of the U.S. and southern Canada. A glimpse of this deer's *"white flag" tail* disappearing into a forest, or of a doe with its *white-spotted fawn* are high points of any day's outing. Although the Whitetail is *reddish* most of the year, its coat becomes *grayish* in winter. It eats twigs, grasses, fungi, and acorns. Its excellent sense of smell enables it to pick up scent of humans and move off without being seen. It is the most important big game mammal in the East, but it can become a nuisance in crop fields and orchards. Because its major predators (wolves and cougars) are extinct over much of its range, the Whitetail sometimes becomes overpopulated. Hunting can help control its numbers.

Whitetails are excellent swimmers and often are found on islands in lakes. They are also common in woody suburbs and are often struck by automobiles. Active day or night, they can run up to 35 mph, jump over obstacles 8 ft. high, and cover 30 ft. in one bound. Smaller forms of the Whitetail can be seen in the Florida Keys (**Key Deer**) and in southern Arizona.

The Whitetail's range overlaps that of the Mule Deer (p. 104) from Minnesota west. Where both occur, the Whitetail can be identified by its *smaller ears* and feet, its longer tail *(without any black above)*, a more slender neck, a narrower face, and daintier legs. The males' antlers are low and compact, with several short, *unbranched tines* on 2 main, forward-leaning beams. Males can weigh up to 400 pounds.

winter male

summer female

WHITETAIL DEER

head in velvet: summer

fawn

l.f.

20 in.

3 in.

Whitetail
Deer

r.h.   r.f.

HOOFED MAMMALS: DEER

MULE DEER   To 3½ ft. high (at shoulder)
Best known for its long, *mule-like ears* that
move constantly and independently as it lis-
tens for danger. The Mule Deer differs from
the Whitetail not only by its larger ears, but
also by its *shorter tail,* which is *black
above;* its thicker neck; wider face; and
stockier legs. The antlers of the male grow
higher and each branch divides into equal
forks, unlike a Whitetail's, which have
many unbranched tines. When a Mule Deer
senses danger, it leaps in an odd way called
stotting—stiff-legged, with all 4 feet off the
ground. Males weigh up to 400 pounds.

Typical Mule Deer live in the mountains,
plains, and deserts of the West, from western Texas, Arizona, and California north
through western Canada. These deer have
white rump patches, a *black-tipped tail*
above (white below), and *contrasting* blackish heads, with whiter faces and throats. In
the dense, moist forests of the Pacific Coast,
from northern California to the Alaskan
Panhandle, live 2 smaller races known as
Blacktail Deer. These deer are darker
brown. The top of the tail is *entirely blackish,* not white at the base.

The Mule Deer is the most important big
game mammal of the West. Active day or
night, it may be seen in most western parks
and protected areas. It follows definite
trails, and in mountain areas it will spend
summers in high country and retreat to valleys in winter, sometimes in groups.

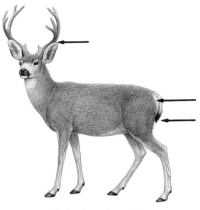

Rocky Mountains: winter

MULE DEER

Northwest Pacific Coast: winter

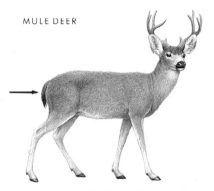

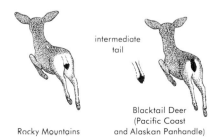

intermediate tail

Rocky Mountains

Blacktail Deer
(Pacific Coast
and Alaskan Panhandle)

HOOFED MAMMALS:
MOOSE AND PRONGHORN

MOOSE To 7 ft. high (at shoulder)
Weighing up to 1400 pounds, the Moose is
the largest deer in the world. It is dark
brown, with *gray legs,* an *overhanging
snout,* a long *dewlap* or beard hanging
from the throat, and a *shoulder hump.*
Males grow *massive, flat antlers* with small
prongs projecting from the borders. Elk
(Wapiti) and Caribou (p. 100) have pale
rumps and lack the long face and snout.
The Moose occurs in *northern forests,*
where it is often seen feeding on water
plants far out in ponds and lakes. In winter
it feeds on twigs, bark, and saplings. It is a
fast swimmer, and on land it can run as
fast as 35 mph. The rust-colored young lack
the white spots of other deer fawns. Found
across Canada and Alaska, the northern
Rockies, upper Midwest, and northern New
England. Best seen in the following parks:
Baxter (Maine), Algonquin, Isle Royale,
Grand Teton, Yellowstone, Glacier, Banff,
Jasper, and Denali.

PRONGHORN To 3 ft. high
The only member of the Pronghorn family,
found only in the arid plains of the American West. Sometimes mistakenly called an
antelope. Best seen in Yellowstone, Wind
Cave (South Dakota), and Pawnee National
Grassland (Colorado). The Pronghorn has
true horns rather than antlers, which are
found only in the deer family. True horns
are not normally shed each year, but the
Pronghorn's horns are covered by horny
sheaths that are *shed each year.* All males
and most females have horns. Both sexes
have conspicuous *white rumps,* white
throat stripes, and white sides. Males alone
have black patches from nose to eyes and on
the neck. The Pronghorn browses on weeds,
sagebrush, and grasses. When alarmed it
raises its white rump hairs and dashes off
at speeds of over 50 mph. It is the fastest
mammal in the Western Hemisphere and
one of the fastest in the world.

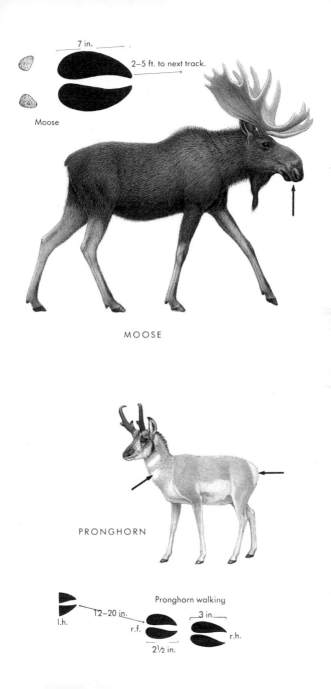

7 in.

2–5 ft. to next track.

Moose

MOOSE

PRONGHORN

Pronghorn walking

l.h.

12–20 in.

r.f.

2½ in.

3 in.

r.h.

HOOFED MAMMALS:
MUSKOX AND BISON
Wild Goats, Sheep, and Relatives (Bovids)

This family includes the Bison, Muskox, goats, sheep, cattle, and the true antelope and buffalo of the Old World. These mammals have *unbranched horns* that are *never shed*. The horns are made up of a sheath covering a bony outgrowth of the skull. Horns are present in *both sexes*, not just males.

MUSKOX To 5 ft. high (at shoulder)
A brown ox of the far northern tundra, with *long, silky* hair that *hangs down* nearly to its feet. The lower legs and back ("saddle") are silvery white. The broad, flat horns are plastered close to the skull, with curved tips that point forward. The female's horns are more slender and curved. Small herds graze in valleys in summer and seek windswept slopes and hilltops with less snow in winter. They usually face danger by forming a circle with their heads facing outward against wolves or human hunters. Once important as a food item for Eskimos, Muskoxen were exterminated over much of their range. They are being reintroduced in parts of Alaska and northern Canada. A Muskox can weigh up to 900 pounds.

BISON To 6 ft. high
This enormous mammal (which can weigh up to 2000 pounds) is dark brown, with a *high hump* on its shoulders that is usually pale golden brown. The *head* is *massive* and both sexes have *horns. Long, shaggy hair* hangs from the shoulders and front legs. Before North America was settled by Europeans, Bison numbered perhaps 60 million, but by 1900 fewer than 1000 remained. A last-minute rescue attempt resulted in a successful comeback. Now you can see Bison at Yellowstone, Wind Cave (South Dakota), Wood Buffalo (Alberta), and many other parks and refuges. Bison are also called Buffalo, but true buffalo live only in Africa and Asia.

r.f.　r.h.

MUSKOX

BISON

Bison

5 in.　3 ft. or less to next track.

HOOFED MAMMALS: GOATS AND SHEEP

MOUNTAIN GOAT To 3½ ft. high (at shoulder)
A yellowish white mammal with a mane on its throat that looks like a beard. Its short, smooth, *thin, black horns* curve slightly backward. In summer its hair becomes shorter. At that time of year it is found on rocky crags near snowline. In winter it descends to lower elevations and develops a longer, shaggier coat. An excellent climber, it has flexible black hooves, a compact and muscular body, and short legs that are ideal for balance but poor for running. The Mountain Goat is famous for making its way along incredibly narrow ledges. Small herds inhabit mountains from Montana to southern Alaska.

BIGHORN SHEEP   To 3½ ft. high
A *thick-necked* sheep with a *creamy white rump.* The Bighorn is larger and dark brown in the northern U.S. and Canadian Rockies and smaller and pale tan in southwestern deserts. Males feature *massive coiled horns* that spiral back, out, and then forward, forming a full *curl.* Females have shorter, thinner horns, resembling those of Mountain Goats. Bighorns live on mountain slopes, in meadows, and in rugged rocky areas with few trees. Males engage in serious butting contests in fall, when the sounds of their crashing heads can be heard over a mile away. Bighorns are best seen at Death Valley, Yellowstone, and Glacier.

WHITE SHEEP   To 3½ ft. high
This wild sheep replaces the Bighorn in Alaska and Yukon. Often known as Dall's Sheep, it is *all white* with black hooves and massive yellowish horns that are somewhat *more slender* than those of Bighorns. It is common at Denali Park in Alaska. In southeastern Yukon and northern British Columbia a dark form known as **Stone's Sheep** occurs. This dark form varies from black to brown or silver, with a *contrasting* white rump, belly, face, and *leg trim.*

Mountain Goat

About 15 in. to next track.

3 in.

MOUNTAIN GOAT

3 in.

About 15 in. to next track.

Bighorn Sheep

BIGHORN
SHEEP

black phase

white phase

WHITE SHEEP

HOOFED MAMMALS

PECCARY (JAVELINA) To 36 in. long

This *pig-like* mammal belongs to the fourth family of hoofed mammals in North America. Unlike the Wild Boar, which was introduced from Europe, this wild pig is native to North and Latin America. The Peccary lives in bands of up to 25 in semi-arid deserts and hills covered with oaks, chaparral, and mesquite in southern Arizona and Texas. It feeds on prickly pear cactus (spines and all), mesquite beans, tubers, leafy plants, toads, lizards, baby birds, and even snakes.

The **Wild Boar** (not shown) has been introduced as a game mammal in California and much of the South. Its tusks are *curved* and curl *out and up*, while the Peccary's tusks are *straight* and *point downward*. The Wild Boar has a long, straight tail.

SIRENIANS

MANATEE To 13 ft. long

The only sirenian (p. 9) in North America. A grayish, *nearly hairless* aquatic mammal with flippers and a broad, *paddle-shaped tail.* Its broad head, with split upper lips and numerous stiff bristles, is all that is usually seen above water. Manatees often hiss when they meet and embrace each other with their flippers. These mammals live in warm rivers and along the Atlantic and Gulf coasts, from North Carolina to Texas. Most are found in Florida. Almost all manatees are seen with scars from motorboat propellers.

EDENTATES

ARMADILLO To 33 in. long

The only edentate (p. 9) in North America. It is covered with *heavy, bony armor* over the head, body, and tail. Flexible bands across the middle of its body allow it to twist and turn. Strong claws enable it to dig burrows in sandy soils or streamsides. The Armadillo feeds chiefly on insects, crayfish, frogs, bird eggs, and berries. Its range has *expanded* recently, from Texas east to Florida and north to Missouri.

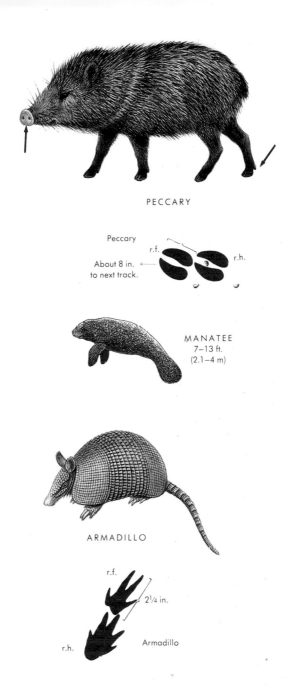

PECCARY

Peccary

r.f.

About 8 in.
to next track.

r.h.

MANATEE
7–13 ft.
(2.1–4 m)

ARMADILLO

r.f.

2¼ in.

r.h.

Armadillo

TOOTHED WHALES

This group is made up of beaked and sperm whales and the dolphins and porpoises. All have *simple, peglike teeth* and no baleen. (Compare with baleen whales, p. 122.)

Beaked Whale Family

Primitive whales, rarely seen and not well known. These whales hunt squid and deep-sea fishes. They have a *narrow snout*, usually with *only 2 or 4* functional teeth. Note the small dorsal fin *near the rear* and the 2 throat grooves.

BAIRD BEAKED WHALE　　　　To 42 ft. long
A large, blackish whale with a whitish area on the lower belly. Has a *long beak*, with the lower jaw protruding beyond the upper. Occurs in schools of up to 30 off the Pacific Coast.

TRUE BEAKED WHALE　　　　To 17 ft. long
One of 7 closely related, little-known whale species. This species is *slate black* above and paler below. The male has *2 small teeth* at the *tip* of the lower jaw. It is found in the North Atlantic, sometimes as far south as Florida.

GOOSEBEAK WHALE　　　　To 28 ft. long
Also known as the Cuvier Whale. Varies in color, often with white patches on the head, back, or belly and a background color of black, brown, or gray. Has a thick body with a *distinct keel* running down the back from the dorsal fin to the tail. The snout is *roundish*, unlike the longer snouts of other beaked whales. This whale lives off both coasts. It swims in groups of up to 40. Males have 2 teeth in the lower jaw.

BOTTLENOSE WHALE　　　　To 30 ft. long
Found only in the cold waters of the Arctic and North Atlantic oceans. Males have a *high forehead* that rises above a short beak and 2 small teeth at the tip of the lower jaw. Older males have a white dorsal fin and a white patch on the forehead. Body color varies from grayish black to light brown, with a whitish belly. The Pilot Whale (p. 120) is similar, but *black overall*, with a *larger dorsal fin* on the *front* half of the body.

BAIRD BEAKED WHALE
35–42 ft.
(10.7–12.8 m)

TRUE BEAKED WHALE
15–17 ft.
(4.5–5.2 m)

GOOSEBEAK WHALE
18–28 ft.
(5.5–8.5 m)

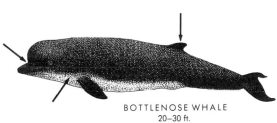

BOTTLENOSE WHALE
20–30 ft.
(6.1–9.1 m)

TOOTHED WHALES

SPERM WHALE
To 60 ft. long

This whale, featured in the book *Moby Dick*, is the only species in its family. It has an enormously high, *squared-off forehead*. Its head is one-third of its length. There are up to 28 strong teeth on each side of the *long, narrow lower jaw*. The back has *no dorsal fin*. When this whale exhales or "blows" at the surface, its spout of mist is *directed forward*. Able to dive down to 2 miles below the surface, the Sperm Whale feeds on giant squid, octopus, bottom-dwelling sharks and other fish. Found off both coasts, it has been overhunted for its oil, which is used to lubricate machines.

PYGMY SPERM WHALE
To 13 ft. long

This whale and the related **Dwarf Sperm Whale** (not shown) are the only small whales with a *forehead (snout)* that *protrudes* in front of the mouth. The Pygmy is black above the grayish white below, with a pale, *bracket-shaped mark* behind each eye. It has a small dorsal fin. The lower jaw is narrow, with many needle-like teeth. Found in warm waters off both coasts.

WHITE WHALE (BELUGA)
To 14 ft. long

This is a small *white* whale of cold arctic waters that feeds on fish and cuttlefish. Note its fairly *high* forehead, *very short snout*, and *lack of a dorsal fin*. It feeds in shallow waters and rivers from Alaska to Hudson's Bay and south to the St. Lawrence River. Young are born brown and gradually turn gray, then white by their sixth year.

NARWHAL
To 15 ft. long

Both sexes are *mottled brownish above* and paler below. They have *no snout* and *no dorsal fin*. Narwhals live in the high Arctic, in cold seas from northern Alaska to Greenland. They feed on fish and deepwater squid and crustaceans. Both sexes have only 2 upper front teeth. In males the left one grows into a long, hollow, twisted, *unicorn-like tusk* that can reach 9 ft. long.

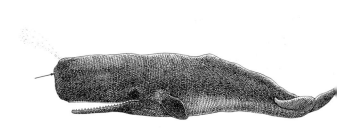

SPERM WHALE
40–60 ft.
(12.2–18.3 m)

PYGMY SPERM WHALE
9–13 ft.
(2.7–4 m)

WHITE WHALE
11–14 ft.
(3.4–4.3 m)

NARWHAL
12 ft.
(3.6 m)

WHALES: DOLPHINS

These small whales usually have a *well-developed dorsal fin* and notched tail flukes. They commonly travel in large groups and often ride the bow waves of ships.

SPOTTED DOLPHIN To 7 ft. long

Fairly common in the Gulf of Mexico and northward to North Carolina. Numerous *large white spots* on a blackish back. The long snout is separated from the forehead by a distinct groove. The **Striped Dolphin** (not shown) is black above and white below, with *black stripes* from eye to flipper and undersides. It is found on both coasts in colder waters.

COMMON DOLPHIN To 8½ ft. long

Found off both coasts, this dolphin is black above with black flippers, contrasting with its *yellow flanks* and white belly. Note the *long beak* and black "spectacles."

PACIFIC BOTTLENOSE DOLPHIN To 12 ft. long

Large and uniformly *grayish*, with a paler belly and a short beak with a slightly longer lower jaw. This Californian species has *white* on the *upper lip*. The **Atlantic Bottlenose Dolphin** (not shown), *common* along the Atlantic Coast, is similar but has *no white* on the upper lip.

RIGHT WHALE DOLPHIN To 8 ft. long

Small and black, with a distinct *white belly stripe* from breast to tail and *no dorsal fin*. Lives in the Pacific from the Bering Sea south.

PACIFIC WHITE-SIDED DOLPHIN To 9 ft. long

Greenish black above, with *whitish sides* and belly. *Blunt-nosed*—lacks the pointed beak and pale face of the Common Dolphin. Found all along the Pacific Coast. The similar **Atlantic White-sided Dolphin** (not shown) is found on the Atlantic Coast south to Cape Cod.

WHITEBEAK DOLPHIN To 10 ft. long

A black dolphin with a *pale flank patch* and a *short white beak*. North Atlantic west to Labrador.

SPOTTED DOLPHIN
5–7 ft.
(1.5–2.1 m)

COMMON DOLPHIN
6½–8½ ft.
(1.9–2.6 m)

PACIFIC BOTTLENOSE DOLPHIN
10–12 ft.
(3–3.6 m)

RIGHT WHALE DOLPHIN
5–8 ft.
(1.5–2.4 m)

PACIFIC WHITE-SIDED DOLPHIN
7–9 ft.
(2.1–2.7 m)

WHITEBEAK DOLPHIN
7–10 ft.
(2.1–3 m)

WHALES: DOLPHINS

KILLER WHALE (ORCA)   To 30 ft. long
This huge dolphin is black above and *white below*, with white extending up the sides. Note the *white oval* behind each eye, the *long dorsal fin*, blunt nose, and long flippers. Found on both coasts. It feeds on large fish, but specializes in attacking seals, sea lions, and other whales.

GRAMPUS   To 13 ft. long
A large dolphin with a bulging, *yellowish head*, no beak, and mottled gray, *slender flippers.* Males are blue-gray above, females brownish. This dolphin has a much larger dorsal fin than the Goosebeak Whale. It is found off both coasts.

FALSE KILLER   To 18 ft. long
A large, *all-black* dolphin. Note its *rounded head, large teeth* on both jaws, and relatively *small dorsal fin.* Travels in very large schools that sometimes strand themselves on beaches. Found on both coasts, from North Carolina and Washington south.

COMMON PILOT WHALE   To 28 ft. long
("BLACKFISH")
A huge, *all-black* dolphin with a *bulging forehead* and a *large, swept-back dorsal fin.* It travels in large schools in the Atlantic as far south as Virginia.

HARBOR PORPOISE   To 6 ft. long
A small dolphin with a *triangular dorsal fin.* It is found all along the Pacific Coast and on the Atlantic Coast south to New Jersey. Common close to shore and in harbors. Note its round mouth, black back, *pink sides*, white belly, and the *dark line* between mouth and flipper.

DALL PORPOISE   To 6 ft. long
All-black, with a striking *white area* on the sides and belly. Note its blunt head and triangular dorsal fin. Found on the Pacific Coast from Alaska south to California.

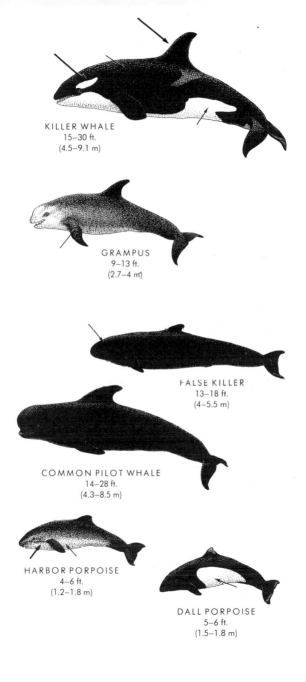

KILLER WHALE
15–30 ft.
(4.5–9.1 m)

GRAMPUS
9–13 ft.
(2.7–4 m)

FALSE KILLER
13–18 ft.
(4–5.5 m)

COMMON PILOT WHALE
14–28 ft.
(4.3–8.5 m)

HARBOR PORPOISE
4–6 ft.
(1.2–1.8 m)

DALL PORPOISE
5–6 ft.
(1.5–1.8 m)

BALEEN WHALES

Large whales with *no teeth* but with *strips of baleen*, or "whalebone," hanging from the edges of the upper jaw. These comblike strips are fringed on the inside and are used to strain food from sea water. When a mouthful of water is forced through the baleen, zooplankton and other small animals are trapped on the fringes and then swallowed. Baleen whales have 2 blowholes. Some species have been hunted to near-extinction.

GRAY WHALE   To 45 ft. long

The only member of its family. A medium-sized, blotched, grayish black whale. It has a series of *bumps on its back*, but *no dorsal fin*. Unlike finbacks (below), it feeds close to shore among rocks and kelp. The Gray Whale raises its young in shallow lagoons off Baja California and western Mexico and migrates north to the Arctic Ocean near Alaska for the summer. Its annual migrations along the California coast attract thousands of whalewatchers.

Finback Whale Family

These whales have a *small dorsal fin* far back on the body and many long grooves on the throat. (Gray Whales have only 2–4.)

HUMPBACK WHALE   To 50 ft. long

A chunky black whale with a white underside. It has *very long, mostly white flippers* with knobs along the leading edges. The head is adorned with *fleshy knobs* and white barnacles. This whale has been severely overhunted, but it can still be seen off both coasts. Lucky whalewatchers have seen the spectacular leaps it makes when breaching and slamming the surface. Known for its lovely, haunting songs.

MINKE (PIKED) WHALE   To 30 ft. long

A small finback whale, found along both coasts. Blue-gray above and white below, with a distinctive *white patch* on its flippers. Its dorsal fin has a *curved tip*. Often approaches boats, sometimes closely enough that its *white baleen* may be seen.

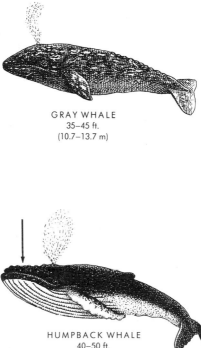

GRAY WHALE
35–45 ft.
(10.7–13.7 m)

HUMPBACK WHALE
40–50 ft.
(12.2–15.2 m)

MINKE WHALE
20–30 ft.
(6.1–9.1 m)

BALEEN WHALES

SEI WHALE (RORQUAL)   To 50 ft. long
This flat-headed baleen whale is larger than a Minke and smaller than a Finback. It is dark above and has a relatively *large dorsal fin*. Its tail flukes are *all dark*, not white on the underside, as in a Finback. Its *baleen* is *black* with white, frayed edges. Unlike the Minke, it has no white patches on the upper side of its flippers. The Sei is found off both coasts. It migrates south to warmer waters for the winter.

FINBACK (FIN) WHALE   To 75 ft. long
A very large, flat-headed baleen whale. It is paler gray above than the Sei and the undersides of its flippers and tail flukes are *white.* The baleen is *streaked purple and white* and the mouth is unique—the right side of the lower lip is white and the left is gray. The Finback is an extremely fast swimmer. Unlike the Humpback, this whale only rarely leaps clear of the water. It is most common in subarctic waters but ranges southward along both coasts, usually well offshore and often in groups.

BLUE WHALE   To 100 ft. long
Weighing up to 200 tons, this whale is the largest animal to have ever lived, dwarfing even the largest dinosaurs. It is *blue-gray* above, with *yellowish white* on the belly. It has a small dorsal fin, white on the undersides of the flippers, and *black baleen.* 1500 Blue Whales summer in the Gulf of Alaska and the Aleutians, migrating south to waters off western Mexico for the winter. These behemoths are seriously endangered. The few hundred survivors off our East Coast summer from the Gulf of St. Lawrence north along Davis Strait, wintering at unknown places in the tropical Atlantic. Blue Whales feed chiefly on shrimp-like krill, strained from sea water by their comb-like baleen.

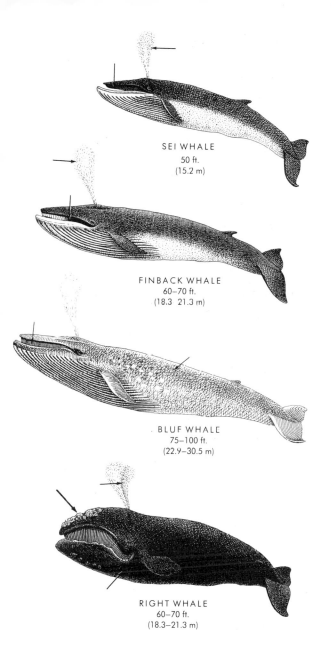

SEI WHALE
50 ft.
(15.2 m)

FINBACK WHALE
60–70 ft.
(18.3 21.3 m)

BLUF WHALE
75–100 ft.
(22.9–30.5 m)

RIGHT WHALE
60–70 ft.
(18.3–21.3 m)

BALEEN WHALES

Right Whale Family

Early whalers named these the "right" ones to kill because their oil-rich bodies would float after they were harpooned. They have an *enormous head*. The arched mouth has *several rows of baleen*, with a gap at the snout. *No dorsal fin.*

RIGHT WHALE To 60 ft. long

A large, blackish whale, with raised, *wart-like knobs* on its head and sometimes with *pale patches* on its underparts. Only a thousand individuals survive off both coasts. The Atlantic population migrates north to the Gulf of St. Lawrence for the summer, the Pacific whales to southern Alaska. Both groups migrate to warmer waters for the winter. The Right Whale's *spout* resembles a V, forming 2 columns up to 15 ft. high.

BOWHEAD WHALE To 60 ft. long

Replaces the Right Whale in cold arctic seas from western Alaska through northern Canada to the Gulf of St. Lawrence. Usually found near the edge of a polar ice field. The *huge head* has a *bowed lower jaw*, which is *mostly white*. Bowheads remain in arctic waters year 'round and are able to break through ice several feet thick.

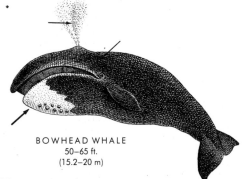

BOWHEAD WHALE
50–65 ft.
(15.2–20 m)